Paumanok Publications, Inc. July/August 2000

Passive Component Industry

An affiliate publication of the
A sector of the Electronic Industries Alliance

The Only Magazine Dedicated Exclusively To The Worldwide Passive Electronic Components Industry

Aluminum Electrolytics

Markets & Competition

Anode and Cathode Foil
Market Dynamics of Thin Foil, Etching and Winding

Organic Polymers
Trends and Directions

Don't let this part
Stop You From Producing These

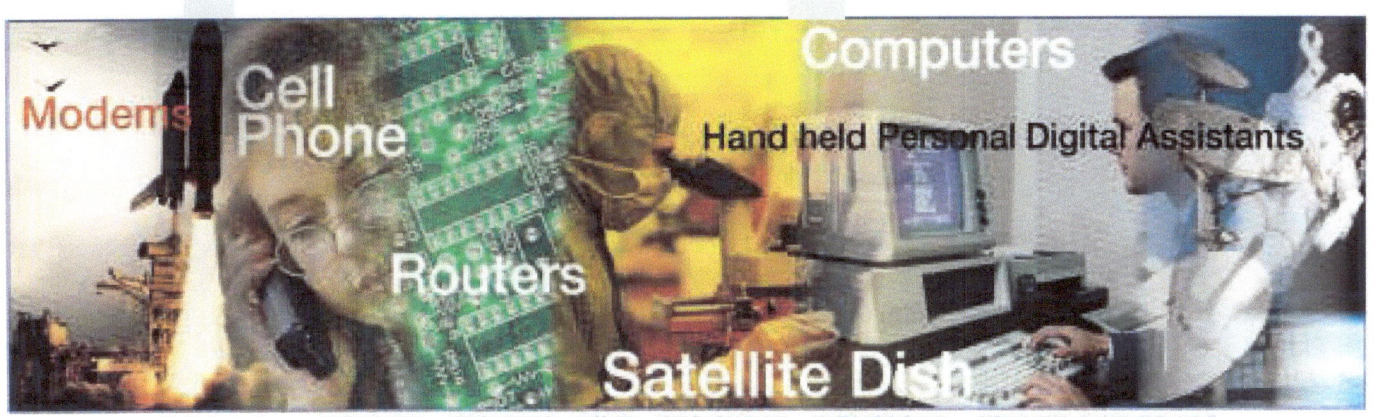

Modems · Cell Phone · Routers · Computers · Hand held Personal Digital Assistants · Satellite Dish

ALUMINUM · TANTALUM · CERAMIC · FILM · RESISTORS · INDUCTORS · FERRITE CHIP BEADS
THERMISTORS · VARISTORS · DIODES

Contact NIC today for alternative component solutions to today's hard a olbiam passive com onelDts...

NIC COMPONENTS CORP.
70 Maxess Road
Melville, New York 11747
TEL (631) 396-7500 / FAX (631) 396-7575

www.niccomp.com

STEINERFILM ® a step ahead

Development and manufacture of non-metallized, metallized and coated capacitor films, capacitor papers and specialty films

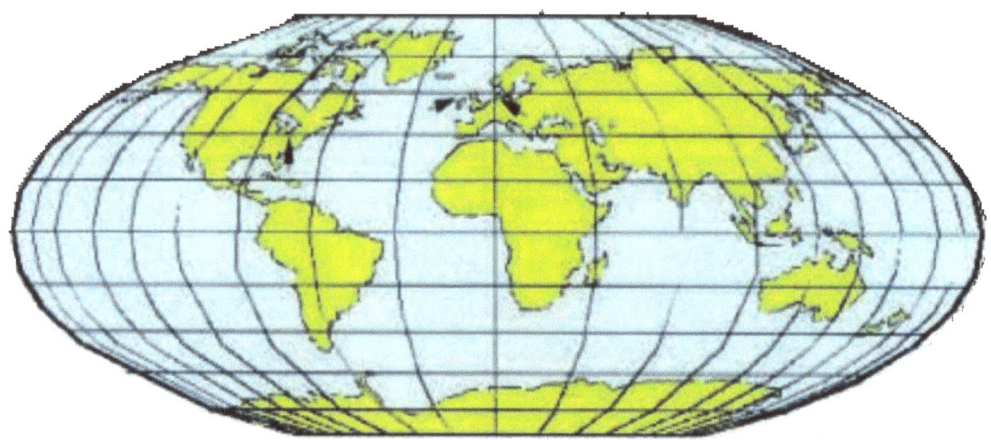

↑ **Manufacturing facilities**

Customer Representatives for the STEINER-GROUP of companies:

Germany:
(since 1951)
STEINER GmbH & Co KG
P.O. Box 70
D–57335 Erndtebrück
Germany
Tel.: +49 2753 6070
Fax: +49 2753 607 153
STEINER@STEINERFILM.DE

Ireland:
(since 1969)
STEINER (Galway) Ltd.
Carnmore, Galway
Ireland
Tel.: +353 91 755444
Fax: +353 91 757722

North–South America:
(since 1971)
STEINERFILM, INC
987 Simonds Road
Williamstown, Mass.
02167 USA
Tel.: +1 413 458 9525
Fax: +1 413 458 2495

Italy:
Transit S.R.L.
Mr. F. Valerio
Via C. Scalzi 20
I–37122 Verona
Italy
Tel.: +39 045 8000196
Fax: +39 045 597512
TRANSIT@SIS.IT

France:
Pierre Foulon
140, Allee de Beauregard
F–75540 Viuz-La-Chiesaz
France
Tel.: +33 450 775505
Fax: +33 450 775649
FOUSTEI@NWC.FR

East-Europe, Scandinavia:
Klesper Chemiehandel
Mr. J Klesper
Am Osterberg 7
D–21266 Jesteburg
Germany
Tel.: +49 4183 3847
Fax: +49 4183 5214
J.KLESPER@T-ONLINE.DE

Great Britain:
Mr. N. Pomery
150, Lightwood Road
Buxton
Derbyshire SK 17 6RW
United Kingdom
Tel.: +44 1 298 77814
Fax: +44 1 298 79943
POMERY@GEMSOFT.CO.UK

Korea:
Mr. Chang Jae Cho
#301 Westvil
370-22, Seokyo-Dong
Mapo-Ku, Seoul
South Korea
Tel.: +82 2 333 4906
Fax: +82 2 333 4908
CJCHO@ELIM.NET

Pakistan:
M/S–AS Trading Co.
3/37 Chaman Chambers
Chowk Dalgram, Lahore
Pakistan
Tel.: +92 42 7663784
Fax: +92 42 7662274
MSARWAR@BRAIN.NET.PK

Malaysia, Thailand, Indonesia, Singapore:
Oriental Marketing
Mr. Lee Yuen Quee
49, Jalan Budiman 26
56100 Bandar Tun Razak
Kuala Lumpur, Malaysia
Tel.: +60 3 9731103
Fax: +60 3 9731108
ORIENTL@PD.JARING.MY

Japan:
Mr. A. Nakao
17-4, 34, 5-Chome Nagayama,
Tama City
Tokio
Japan
Tel.: +81 423 371018
Fax: +81 423 391961

Taiwan:
King Star Enterprises Corp.
Mr. C. Huang
6Fl No.3 Li Shiu St.
POB 84,249
Taipeh/Taiwan
Tel.: +88 62 2391 2578
Fax: +88 62 2393 0087
KSHWCAP@MAIL.SYSNET.NET.TW

China:
Honkison Trading Company
Mr. B. Leung
Block 2, 11th Floor
Wah Shing Centre
11-13 Shing Yip Street
Kwun Tong, Kowloon
Hong Kong
Tel.: +852 2 7930111
Fax: +852 2 27930109
CO@HONKISON.COM.HK

Australia:
Horst Stürmann
1 Drysdale Place
Kareela, N.S.W. 2232
Sydney
Australia
Tel.: +61 2 9521 6486
Fax: +61 2 9545 1047

Spain:
Mr. Juan Staib, S.A.
Pje Dos de Mayo, 3 bjs.
E–08041 Barcelona,
Spain
Tel.: +34 93 4564500
Fax: +34 93 4330580
STAIB@MDIN.ES

India:
Mr. N. P. Trivedi
TRIVTECH Corporation
326, T.V. Industrial Estate
Bombay 400 025
India
Tel.: +91 22 4938403
Fax: +91 22 4930191
TRIVTECH@VSNL.COM

For component metallizations, Heraeus stands behind you...

and ahead of the curve.

For as long as you've been building passive components, Heraeus has supplied metallization products and technology to advance your products and processes. We provide support at every step: new concepts, technical expertise, on-site service, and worldwide leadership. And we continue to stand behind the passive component industry because, frankly, we value your business.

At the same time, we're out in front of the market, with a focused effort to support the shift from precious metals to base metals technology (BME). After a century of experience in precious metals, we're applying our wealth of knowledge to create new products and processes that will propel your company into the forefront of your marketplace.

Looking further ahead, Heraeus is working on the connectivity between end termination and solder paste technology, with an emphasis on lead-free assembly. As the only supplier of products to both the component and assembly industries... and as a leader in the development of the most promising lead-free alloy... Heraeus offers expertise... and synergy.

So whatever your metallization needs may be... from immediate support to future developments... only one source supplies it all: Heraeus.

Expect more from Heraeus.

24 Union Hill Road West Conshohocken PA 19428
Tel: 610-825-6050 • Fax: 610-825-7061 Visit us on the Web at: www.4hcd.com

TABLE OF CONTENTS

Passive Component Industry

MAY/JUNE 2000  Volume 2, No. 4

The Only Magazine Dedicated Exclusively To The Worldwide Passive Electronic Components Industry

FEATURE STORIES

7 Aluminum Electrolytic Capacitors
A close-up look at the markets and technologies in aluminum electrolytic capacitors for 2000.

10 Company Profile: NIC Components
An inside look at the Long Island aluminum electrolytic powerhouse.

13 Organic Polymers: Trends and Directions
A brief look at the trend toward solid aluminum, low ESR capacitors with organic electrolytes.

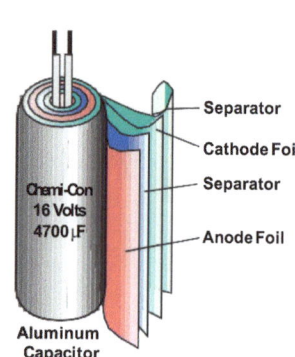

26 Anode and Cathode Foil: Market Dynamics of Thin Foil, Etching and Winding A look at the supply process for thin, high-purity foil to anode and cathode foil etchers before aluminum capacitor winding.

DEPARTMENTS

5 Letter From the Publisher
Recent price increases of ruthenium add pressure to the components industry.

14 Featured Technical Paper
Electrolytes for High Voltage Aluminum Electrolytic Capacitors.

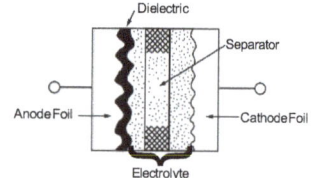

27 Events
Inside Ferro's open house and the ECA summer conference.

32 Market Statistics
Changing markets in passive component distribution: 1999-2005.

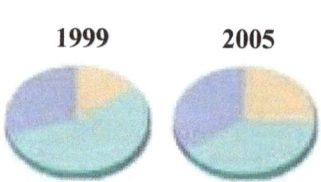

33 Newsmakers
New product offerings and important developments in the passive component industry.

Cover Art: Aluminum Electrolytic Capacitors, courtesy NIC Components.

Letter From the Publisher

PUBLISHER
DENNIS M. ZOGBI

BUSINESS MANAGER
DIRECTOR OF ADVERTISING
SAM COREY

EDITOR
PAMELA GABRIEL

MARKETING
STEPHEN PETTEWAY

ART DIRECTOR
AMY DEMSKO

EDITORIAL ADVISORY BOARD

James M. Wilson
Murata Electronics N.A. Inc.

Glyndwr Smith
Vishay Intertechnology, Inc.

Ian Clelland
ITW-Paktron

Pat Wastal
Avnet

Craig Hunter
AVX Corporation

Jeff Kalb
California Micro Devices

Daniel F. Persico Ph.D.
Kemet Electronics Corporation

Editorial and Advertising Office
109 Kilmayne Drive, Suite A
Cary, North Carolina 27511
(919) 468-0384   (919) 468-0386 Fax

The Electronic Components – Assemblies – Materials – Supplies Association (ECA) represents the electronics industry sector comprised of manufacturers and suppliers of passive and active electronic components, component arrays and assemblies, and commercial and industrial electronic component materials and supplies. ECA, a sector of the Electronic Industries Alliance, provides companies with a dynamic link into a network of programs and activities offering business and technical information; market research, trends and analysis; access to industry and government leaders; standards development; technical and educational training; and more.
The Electronic Industries Alliance (EIA) is a federation of associations and sectors operating in the most competitive and innovative industry in existence. Comprised of over 2,100 members, EIA represents 80% of the $550 billion U.S. electronics industry. EIA member and sector associations represent telecommunications, consumer electronics, components, government electronics, semiconductor standards, as well as other vital areas of the U.S. electronics industry. EIA connects the industries that define the digital age.
ECA members receive a 15% advertising discount for *Passive Component Industry*. For membership information, contact ECA at (703) 907-7536 or www.eia.org/cg, contact EIA at (703) 907-7500 or www.eia.org.

The global market for passive components continues to remain extremely robust in July of 2000 as demand continues to exceed supply for many different types of products, with emphasis upon tantalum chip capacitors, MLCC, film chip capacitors, surface mount chip resistors and chip inductors. The primary reason for the supply/demand imbalance is the explosive growth in demand for cellular telephones, which were forecast to increase from 283 million units shipped globally in 1999 to 420 million units shipped in 2000. In July, however, Paumanok Publications, Inc. received a report from some of its customers suggesting that Nokia had downgraded its global production estimate for Nokia cellular phone production from 180 million phones to 150 million phones. This, in turn, alleviated some of the supply constraints on P case size tantalum chip capacitors coming out of Japan.

Other companies directly involved in selling MLCC note their fears are centered around the fact that business is too good; shipments of MLCC are so great in 2000 that they cannot all actually be consumed, and that some must be stockpiled in inventories, which, in turn, will cause a significant downturn in MLCC demand sometime at the end of the first quarter of 2001.

Other recent developments of extreme importance include the tremendous increase in price of ruthenium metal, which is used in thick films for the majority of chip resistors produced worldwide. In fact, the price of ruthenium has increased from $51.00 per troy ounce in January of 2000 to $160.00 per troy ounce in July of 2000. This represents an increase of 215% in price in just seven months. The world's ruthenium supply comes primarily from the Rustenberg Platinum Mine in South Africa, which is showing a rapid increase in demand for the metal for applications in superalloy flange and fitting production. This puts added pressure on the electronic components industry, which consumed 50% of the global mine output of ruthenium in 1999. Due to the heavy reliance on the South African mine, coupled with the growth in chip resistor sales and the new-found uses for ruthenium, it is expected that prices for ruthenium will continue to increase throughout the year.

Dennis M. Zogbi

The other topic of great importance are the rumors of long term agreement contracts currently being signed in the passive component industry, especially for tantalum capacitors. These contracts offer three- and five-year set rates of price erosion in the 5% to 8% range per year. It has been reported that both original equipment manufacturers, especially those producing wireless communications devices, and a few contract electronic manufacturers are entering into long term agreements with major suppliers of tantalum and ceramic chip capacitors to guarantee near-term shipments of components. The danger of such a relationship between capacitor manufacturer and customer is that, should prices drop drastically in the second quarter of 2001 due to overexpansion and prices drop beyond 5% or 8% (historically, they have dropped by as much as 20% in excess sup-

Continued on page 24

Admit it.

You know you should be thinking more about future trends and what they mean for your company. Here's the problem:

You don't have time.

What you need is a concierge of revolutionary thinking, an infomediary for theories of change...Someone who will condense the best ideas on social trends and the future of business into bite-sized, easily digestible packages and serve them up in a convenient format...Someone who will point you to the best thinkers, the best articles, the best books from among the hundreds competing for your scarce attention...Someone who can offer customized products to allow your staff to interact in proprietary ways and develop collaborative ideas for your company online.

The solution is here.

TrendScope—an electronic publication of the Electronic Industries Alliance, the group representing the industries leading us into a new economy and a new era.

Issues arrive on your desktop once a month packed with informative, entertaining, and mercifully brief and punchy features. You can put everyone in your organization on our mailing list. Every last one. FREE. We'll even set up password-protected discussion sites and bulletin boards just for you, so you can use TrendScope as an engine of change within your walls.

Subscribe yourself and others at www.eia.org/trendscope or email trends@eia.org. Visit our Web site at www.eia.org/trends.

You may as well start thinking about the future. It's here.

FEATURE

Aluminum Electrolytic Capacitors: Market Overview and Description

Aluminum electrolytic capacitors, while not as volumetrically efficient as other capacitors, offer the lowest cost per microfarad of the four primary dielectrics—ceramic, tantalum, aluminum and DC film.

In the past, aluminum electrolytic capacitors have been known for their limited shelf life and poor low-temperature characteristics. Because of this, they are traditionally used in short shelf life electronic equipment, with emphasis upon consumer audio and video imaging products. In recent years, however, significant improvements have been realized, and both of these historical shortcomings have been improved upon, so that aluminum capacitors are now used in applications that also include computer printed circuit boards, modem cards, battery chargers for cellular phones, PDAs, air bag circuits, lighting ballasts and medical electronic devices.

Aluminum Electrolytic Capacitors

Market Size and Competition

Paumanok Publications, Inc. estimates that in 2000, the global market for aluminum electrolytic capacitors will be worth approximately $3.3 billion USD, with approximately 80 billion pieces consumed. Historically, production of aluminum electrolytic capacitors developed in conjunction with increased production of consumer audio and video imaging equipment, with emphasis upon television sets, computer monitors, stereo equipment and modern telephones—products which are produced primarily in Asia. In 2000, it is estimated that 90% of global aluminum electrolytic capacitor production will occur in Japan, China, Taiwan and the other countries in the Asian region outside of Japan.

The major global manufacturers of aluminum electrolytic capacitors are Japanese, and include Nippon Chemi-Con, Nichicon, Rubycon and Panasonic. In Europe and the United States, aluminum electrolytic capacitor production is limited and centered around the supply of aluminum capacitors for computer terminals and for specialty motor start product lines. European manufacturers of aluminum capacitors include BC Components (Belgium), BHC Aerovox (UK), EPCOS (Germany) and Vishay-Roederstein (Portugal). U.S. producers of aluminum capacitors are also quite limited, with Cornell Dubilier (South Carolina), BC Components (South Carolina) and Nippon Chemi-Con (UCC-Michigan) the major domestic manufacturers.

Applications

Aluminum electrolytic capacitors are used for four

Aluminum Electrolytics

primary functions: (1) general purpose smoothing of signals (ripple current), which is by far the largest application; (2) energy storage (DC bus), which can be considered an extension of general purpose smoothing because of its application in power supplies; (3) pulse applications, which release large bursts of energy in a short period of time (e.g., flash cameras, strobes and implantable defibrillators); and (4) specialty applications, such as motor start, which are generally short-term AC and are largely an extension of pulse applications (they are considered separately because of their high capacitance and voltage requirements).

In most real-world applications where volumetric efficiency is not an issue and where high capacitance is required, aluminum electrolytic capacitors are used in bulk. This is most apparent in consumer audio, video and voice equipment, with emphasis on television sets, stereos and conventional telephones, where aluminum electrolytic capacitors are used quite frequently for general purpose smoothing applications on the printed circuit board.

The most profitable applications for aluminum electrolytic capacitors are in the output filters for power supplies (where many large can aluminum electrolytic capacitors in the 2,200, 3,300 and 4,700 µF range are employed in series) and in the specialty pulse discharge and photoflash aluminum capacitor markets (which combine high voltage and high capacitance, traditionally an aluminum electrolytic capacitor solution or a power film capacitor solution). These specialty and niche market applications include motor start capacitors (e.g., garage door openers), implantable defibrillators (not external defibrillators, which generate their pulse using a power film capacitor), camera flash, electronic ballasts and air bag igniters.

Capacitance Range and Dielectric Competition

Aluminum electrolytic capacitors are applicable in the capacitance range from 0.47 µF to 1,500,000 µF (1.5 farads). In this capacitance range, aluminum electrolytics compete with solid tantalum capacitors, which are applicable in capacitance ranges from 0.0047 µF to 1,200 µF. Tantalum capacitors are more expensive per µF but have better volumetric efficiency, and therefore are chosen over aluminum electrolytic in most handheld applications, such as cellular phones and pagers. In the first and second quarters of 2000, however, due to the shortage of available tantalum chip capacitors, many tantalum chip capacitor customers opted to purchase surface mount aluminum electrolytic capacitors to satisfy production quotas. This SMD aluminum for tantalum displacement event was most prevalent in the larger capacitance applications (e.g., 100 µF) in the automotive electronic subassembly, computer and telecom infrastructure equipment markets.

In the second quarter of 2000, it has also been reported in Japan that the supply of surface mount aluminum electrolytic capacitors is becoming constrained, not only because of their use in replacing substantially hard-to-find tantalum chip capacitors but also because of substantially increased demand from the Japanese DVD player and Playstation markets. A DVD drive contains between 20 and 100 SMT aluminum capacitors (in the AV amplifier). Demand is also reported to be high at Sony Corporation for its production of the Sony Playstation (Model 2), which requires 36 conventional SMT aluminum capacitors and an additional 13 solid polymer aluminum capacitors. Sony estimates it will produce up to two million Model 2 Playstations in 2000.

In addition to competition from tantalum capacitors, aluminum electrolytic capacitors also receive additional competition from base-metal electrode multilayered ceramic chip capacitors (BME MLCC) in the 1 µF to 22 µF range. Base metals of nickel and copper displace costly but traditional palladium and palladium-plus-silver electrodes and offer higher capacitance in a standard MLCC package. Once again, however, in the first and second quarters of 2000, the supply of base-metal MLCC was also quite limited, which in turn relaxed the threat of encroachment of BME MLCC against aluminum capacitors, at least in the near-term business cycle.

Recent developments in double-layer carbon and mixed-metal oxide "supercapacitors" also threaten some specialty applications traditionally dominated by aluminum electrolytic capacitors. Supercapacitors have the highest capacitance value per cell and extend the aluminum electrolytic capacitance range into the farad arena. Many traditional aluminum electrolytic capacitor manufacturers, such as Matsushita, Elna and Hitachi, also manufacture double-layer carbon supercapacitors as a logical extension of their high capacitance aluminum capacitor businesses.

Voltage Rating

The majority of aluminum electrolytic capacitors are employed at voltage ratings from 6.3 volts to 50 volts, with particular emphasis on the popular 16-volt parts. However, voltages of motor start aluminum electrolytic capacitors can be as high as 200 V, 450 V and 600 V, so aluminums will generally run the gamut with respect to voltage. One of the unique aspects of aluminum electrolytic capacitors is their ability to offer high capacitance and high voltage in the same package. The only other type of capacitor that can accomplish this is the power film capacitor, but at a much higher price.

Aluminum Electrolytics

Configurations

Aluminum electrolytic capacitors are available in leaded and surface mount configurations. Surface mount aluminum electrolytics are still only a small portion of the market, but are certainly the fastest growth portion of the business. Companies that dominate the supply of surface mount aluminum electrolytics include Nichicon and Nippon Industries through NIC Components. The bulk of sales of aluminum electrolytic capacitors continues to be radial leaded devices. Radial leaded devices are further dichotomized between standard radial, snap-in and screw-terminal. Dual and single-leaded axial designs are also available.

In most consumer electronic applications, capacitance values less than 1,000 µF are generally radial leaded or surface mount in design; capacitance values from 1,000 µF to 4,700 µF are generally of multipin snap-in design; and products with values greater than 4,700 µF are usually screw terminal (computer grade).

Surface mount aluminum electrolytic capacitors typically fall into the 1 µF to 47 µF range and are generally found in large numbers only in personal computers and modem cards, although these parts have been noted in smaller numbers on other computer add-on cards such as video and sound cards for DVD players and consumer entertainment boards.

Construction

The construction of an aluminum electrolytic capacitor requires high-purity aluminum foil (for the anode and the cathode), which is usually supplied by KDK, JCC, Becromal or Satma. Anode foils account for the highest costs associated with producing aluminum electrolytic capacitors. The foil is etched into tunnels that provide a surface area for the formation of aluminum oxide, which provides the capacitance.

Kraft or manila separator paper (or combinations thereof) is usually supplied by NKK or Japan Pulp and Paper and is soaked in an ethyl glycol or specialty electrolyte, which is typically mixed in-house or supplied by Tomyama or Yoneyama Chemical. The paper is then rolled between the anode and cathode layers of aluminum foil. These electrolytic-grade papers are impregnated with an electrolyte (in small amounts because of high costs). Electrolytes are also produced captively by many capacitor manufacturers because they can be doped with proprietary materials that enable a company to produce a unique product line. The foil and impregnated paper layers are rolled with leads or tabs and then placed into an aluminum can. The top of the can is then sealed with a gasket assembly.

Technical Trends

Technical trends in aluminum electrolytic capacitors, like technical trends in other dielectrics, begin at the raw material level; aluminum electrolytic capacitors are no exception. Every year companies make small strides in increasing the capacitance value per cubic centimeter of etched and formed aluminum foil; some focus on the development of electrolytes that offer higher voltages in the finished capacitor without using more raw materials. The latest trend focuses on the development of solid aluminum electrolytic capacitors. These capacitors employ organic polymers such as polypyrole and polythiolene that significantly lower the equivalent series resistance of aluminum capacitors so they can release their charge in a faster manner in high frequency applications such as computer motherboards and related applications. ❑

Company Profile: NIC Components

In 1982, NIC Components Corp. was licensed by Nippon Industries Co., Ltd. of Japan for the North American sales of its passive components. Nippon Industries was founded in 1975 by Yoshiharu Dangi and Gichu Sato with an initial investment into a small manufacturer of aluminum electrolytic capacitors. In the ensuing years, they continued to invest in small and medium-sized Asian makers of passive components. Nippon's philosophy was to provide an export market to those independent factories in return for a long term allocation. To further enhance this unique fabless model, Nippon provided its suppliers with financing, engineering and access to high quality raw material suppliers. All of Nippon's products were for export, and many of its cus- tomers initially were importers and distributors in both Europe and North America.

In 1982, Richard Schuster and several associates founded NIC Components Corp. NIC, headquartered in Farmingdale, New York, set up sales and marketing in the United States and Canada and procured most of its product from Nippon Industries. Some of its early distributors included Future Electronics, PUI, Belford, Capsco, Bell (now Arrow) and Brevan. In 1989, NIC opened its second sales and warehouse facility in San Jose, California. While the core business remained in aluminum electrolytics, new fabs manufacturing tantalum, film and ceramic capacitors were recruited to round out the package of passive com ponents. NIC also ventured into the resistive and magnetic com ponent markets with Nippon's new fab liaisons. In 1997, NIC Eurotech Ltd. was established in the United Kingdom as a wholly-owned subsidiary, and in 1999, NIC Asia PTE Ltd. was established in association with Nippon Industries and local management in Singapore. Today, NIC Components is approaching 200 million dollars in sales and has several thousand active customers worldwide, including top tier CEMs and OEMs. NIC's major distributors are Arrow, Future, Kent, Jaco, PUI, Capsco, Chris, Belford, Brevan, First Phase, Shannon, Priebe, U-Tech and Hammond.

Aluminum Electrolytic Capacitors

NIC's original product line encompassed through-hole construction in axial, radial and snap mount configurations. Most of the product was produced in Japan, but due to labor costs and exchange rate considerations, some production was moved to Taiwan and China in the late 80s. Raw material and engineering were still predominantly Japanese in order to assure quality and uniformity.

In the mid-80s, NIC introduced surface mount cylindrical can aluminum electrolytics to the U.S. market. At first the going was slow, with designs coming primarily from the larger, more progressive OEMs. Surface mount was still relatively new and the larger size aluminums presented some challenges for the pick-and-place equipment. Sizes were not yet standardized, and both CV and characteristics were somewhat limited. Still, NIC had resolved that this was the technology of the future, as was evident from the rapid evolution of through-hole to surface mount in other passive com ponents such as ceramic and tantalum capacitors and discrete resistors. Using both through-hole and surface mount components on the same PCB must utilize two soldering processes, which is very costly. Aluminum electrolytics in surface mount packages have only taken off in the last few years, and now they are truly coming into popularity. Expansive ranges of size, capacitance, voltage and spe- cial characteristics such as low ESR, extended temperature and low leakage current offer a multitude of design options.

Due to NIC's early entry into this technology and the fact that they have the most extensive line of surface mount types, they have established a very strong market position. They are now shipping over 40 million pieces per month, and this number is growing at a frenetic pace.

NIC Aluminum Electrolytic Capacitor Technology and Trends

NIC's Aluminum Electrolytic Capacitor offering currently covers:
15-series: Types of surface mount (SMT) parts
25-series: Types of radial leaded parts
5-series: Types of axial leaded parts
4-series: Types of large can (snap-in) parts

Temperature Ratings
-40 to +85°C: General purpose (lowest cost)
-55 to +105°C: Wide temperature (with 4X longer life than +85°C rated)
-55 to +125°C: Extended temperature (with 4X longer life than +105°C rated)

Capacitance Value and Voltage Ranges
Surface mount (SMT): 0.1 to 6,800 µF; 2.0 to 450 VDC
Radial leaded: 0.1 to 15,000 µF; 6.3 to 450 VDC
Axial leaded: 0.47 to 22,000 µF; 6.3 to 500 VDC
Large can (snap-in) leaded: 56 to 68,000 µF; 10 to 450 VDC

Specialty Types
Low impedance-low ESR styles: 3-series surface mount

(SMT) and 5-series radial leaded for high frequency and high current switching power supplies; DC-DC converters and voltage regulator module applications. NIC has recently expanded the range of its NSP series specialty polymer elec- trolyte (solid aluminum) type in SMT package.

Low Leakage Current Styles
1-series surface mount (SMT) and 3-series radial leaded for leakage current sensitive applications (sensors and battery powered circuits).

Bi-Polar Styles
1-series surface mount (SMT), 3-series radial leaded and 1-series axial leaded for applications where circuit voltage bias is unknown or may reverse.

Technology Trends
Alternates to Chip Tantalum
In today's market, alternatives to tantalum chip technology, such as NIC SMT aluminum electrolytic capacitors, are becoming increasingly attractive to circuit designers and PCB manufacturers. Those users adopting easier-to-obtain SMT aluminum electrolytic capacitors, in place of long lead time tantalum chips, have also found a number of performance and cos t-related advantages. Aluminum electrolytic styles have featured improved immunity to unforeseen reserve voltage and over voltage transient conditions, as com pared to tantalum electrolytic styles. Another nice advantage of aluminum electrolytic capacitors (SMT and leaded) is their relative lower cos t, when compared to tantalum solutions.

Aluminum Electrolytic Capacitor Road Map
Majority of development efforts have continued to focus on:
- SMT format development.
- Expanded range of values (introduction of larger case sizes and improved foils).
- Lower impedance-lower ESR styles for next generation lower voltage-higher current circuit designs.
- Improvement in longer life styles.
- Environmental impact issues (alternatives to PVC insulation sleeves of leaded styles).

Expectations
- SMT style usage should exceed leaded styles within the next two years.
- Further reduced pricing of SMT styles is expected as other producers (outside Japan) enter market.
- Axial leaded styles will continue to fall from usage, being replaced by SMT and radial leaded styles.

NIC Components is well positioned in the North American market for passive electronic components. By maintaining a knowledgeable engineering group, an aggressive sales team, and a dedicated source for passive components, NIC Components will continue to grow rapidly in tier one accounts in telecom, computer and automotive end-use market segments. ❑

Passive Component Industry (ISSN 1527-9170) is published bimonthly by *Paumanok Publications Inc.* 109 Kilmayne Drive, Suite A Cary, North Carolina 27511 USA ©2000 *Paumanok Publications Inc.* All rights reserved. Reproduction in whole or part without written permission is prohibited. POSTMASTER: Send address changes to *Paumanok Publications Inc.* at 109 Kilmayne Drive, Suite A Cary, NC 27511. Annual subscription rates for nonqualified individuals: $65.00, U.S.; $75.00, Mexico; $85.00, Canada; $130.00, other countries. Back issues $25.00 when available.

FEATURE

Organic Polymers: Trends and Directions

Aluminum electrolytic capacitors that employ solid polymer electrolytes, which are popularly referred to as solid polymer aluminum capacitors, have and will continue to receive a significant amount of attention in the passive electronic component industry.

Solid polymer aluminum capacitors, largely popularized by Sanyo Video Components in the form of the OS-CON, are an excellent choice for circuit designs that employ the latest microprocessor-based technology. Such circuits require large energy storage capacitors that exhibit extremely low equivalent series resistance (ESR) for almost instantaneous delivery of current. The additional benefits of solid aluminum capacitors include their ability to handle high surge (in-rush) currents and greater AC ripple current when compared to conventional aluminum electrolytic capacitors and solid tantalum capacitors with manganese-based cathodes.

To achieve lower ESR, solid aluminum capacitors employ an organic semiconductor electrolyte. The initial types of electrolytes employed in solid aluminum capacitors were based upon isoquinolinium, a complex salt structure. More recent developments include the use of polypyrole and polythiolene. Organic semiconductor materials are also used to displace the manganese- based cathode materials employed in solid tantalum capacitors which, in turn, give the solid tantalum capacitors low ESR.

Due to the success of Sanyo's OS-CON in the computer motherboard industry since 1995, many additional companies announced the development of similar solid aluminum capacitors, including Panasonic, Nippon Chemi-Con, Kemet (with Showa-Denko), Nichicon and Japan Carlyt Company. Based upon published reports, it is believed that the global production volume for solid aluminum capacitors will triple to 950 million pieces in 2000, reaching 1.3 billion pieces in 2001. The proliferation of low ESR solid aluminum capacitors among a broader supply base should serve to lower the average price per unit for these devices in 2001. Current pricing averages about $1.00 per piece. ❑

Polymer Aluminum Capacitor Forecasts
1999-2001 (U/MM)

Supplier	1999F	2000F	2001F
Sanyo	130	370	500
Panasonic	120	240	300
NCC	10	190	240
Kemet/Showa	10	100	200
Other	15	50	100
Total	285	950	1,340

"Other" includes Nichicon and Japan Carlyt (Suzuki)
Aluminum only; excludes tantalum polymer capacitors
Source: Compiled from Published Reports

Electrolytes for High Voltage Aluminum Electrolytic Capacitors

Robert S. Alwitt, Boundary Technologies, Inc.
PO Box 622 Northbrook, IL 60065-0622
(847) 480-3044

Yanming Liu
Wilson Great Batch Unlimited

Summary

The demand for aluminum electrolytic capacitors with voltage ratings of 450 V and higher has surged due to new applications in motor drives and power electronics. Capacitor performance at high voltage and temperature is constrained by the availability of electrolytes that can provide reliable service. We briefly review the role of the electrolyte in capacitor operation and show how electrolyte chemistry has evolved to meet new requirements. Test data are presented for capacitors containing a new family of high voltage electrolytes that exhibit excellent parametric stability up to 550 V at 105°C.

Introduction

Open the case of an aluminum electrolytic capacitor, and you find a wound element of aluminum foil and spacer paper that is saturated with an organic electrolyte. The electrolyte is critical to capacitor performance. It determines the operating temperature range and has a major effect on DF, ripple current rating and capacitor lifetime. Improvements in electrolytes have been at the heart of many of the improvements in capacitor performance through the years. New electrolytes are key to successful operation of high voltage capacitors under high ripple, high temperature conditions.

Motor drive control may be the most demanding application today for high voltage capacitors. Capacitors must operate with little voltage derating, with high ripple currents that raise the internal temperature, and are expected to provide reliable performance for many thousands of hours. This calls for a low resistivity, low gassing electrolyte. Large motor drives use a bank of capacitors, and then the ESR of individual capacitors must not change appreciably during the operating life in order to maintain electrical balance.

In this paper, we first present some capacitor basics and briefly describe the evolution of electrolytes for high voltage applications. Then we present information on a new electrolyte family that pushes the state of the art to 550 WV at 105°C. An initial test indicates that 600 V at 85°C is within reach.

Fig. 1 shows a sketch of the wound section of an electrolytic capacitor, illustrating its component parts. The anode is aluminum foil that has been electrochemically etched to attain a high surface area. The dielectric is a thin barrier oxide that is electrochemically deposited over the etched anode foil surface. The positive plate of the capacitor is the metal/oxide interface. To realize full capacitance, the conductor at the other oxide face must be in intimate contact with all the surface. In an aluminum electrolytic capacitor this is accomplished with a liquid contact—an organic electrolyte. The electrolyte-saturated spacer is an ohmic resistance, and the capacitor equivalent series resistance

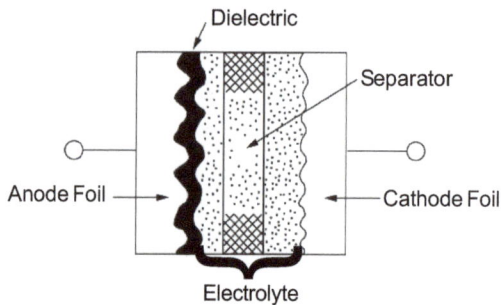

Fig. 1. Basic elements of an electrolytic capacitor.

(ESR) is the sum of the oxide ESR and electrolyte/spacer resistance. The etched cathode foil serves as current collector and is connected to the negative terminal of the package. For high ripple current applications, the cathode capacitance must be large enough to store the ripple charge. The device capacitance is equal to that of the anode and cathode in series.

Electrolyte Characteristics
Properties

An electrolyte is composed of an organic solvent and solutes that provide ionic conductivity. Capacitor electrolytes are formulated to have these properties:
- No breakdown at surge voltage
- Support oxide formation during capacitor aging
- Low resistivity
- Stable properties at maximum operating temperature
- No reaction with Al, Al oxide, or package materials
- Low toxicity and low flammability

Electrolyte Formulations

The basic features of electrolyte formulations are listed here:

1. Ethylene glycol (EG) is the most widely used solvent because it is low cost and provides good electrical properties. Dimethylformamide (DMF) is the solvent in military capacitors which require a very wide operating temperature range. Electrolytes using butyrolactone (BL) as solvent are popular in Asia for low voltage miniature capacitors with low ESR.

2. For high voltage electrolytes with EG solvent, the solute usually consists of ammonium salts of boric acid and/or selected organic acids. Amine salts are used with BL and DMF because ammonium salts have low solubility and are unstable in those solvents.

3. Electrolytes have low water content. Some water is needed to support oxide formation, but too much water causes corrosion of the foil electrodes and generation of hydrogen. A water content of about 3% is suitable for use up to 105°C.

4. An electrolytic capacitor is an electrochemical cell and under applied voltage has a small leakage current. This current generally produces hydrogen at the cathode. Certain chemicals can be added, known as depolarizers, that substitute nongassing reactions for hydrogen generation.

Glycol-borate electrolytes were developed more than 50 years ago and are still in use. Water is a product of the glycol+borate reaction, and the water content of G-B electrolytes is >10%, too high for reliable performance at high temperature. These electrolytes also tend to have high resistivity. To meet today's operating requirements, the borate content must be kept low in order to have a low water content. This means that another solute must be used, either in addition to borate or as a substitute, in order to make a low resistivity electrolyte.

Substitution of an organic acid for boric acid produces electrolytes with lower water content and lower resistivity. Dicarboxylic acids (two acid groups per molecule) perform better than monocarboxylic acids. The straight-chain dicarboxylic acids (DCA) are widely used as chemical intermediates and are available at low cost and high purity. Some members of this family are shown in Fig. 2a. Acid strength decreases with increasing chain length, and weaker acids can be used to higher voltage without breakdown. But solubility decreases with increasing acid molecular weight, and the electrolyte resistivity increases. DDDA is the largest DCA with sufficient solubility in EG to give adequate resistivity. It can be used to 450 V without breakdown.

High voltage electrolytes made with branch-chain dicarboxylic acids (BCAs) are popular in Japan. The structures of some typical BCAs are shown in Fig. 2b. These acids are more soluble in EG than straight-chain acids, and higher molecular weight acids can be used. Performance varies, possibly depending upon details of the acid structure, but some are used to make electrolytes with lower resistivity than DCA electrolytes for the same voltage rating. These capacitors are rated at 400 and 450 WV and have low ESR and high ripple current rating.

Branch-chain acids are used in the chemical industry as cross-linking agent for synthetic resins and rubbers, and their esters are used in cosmetics, lubricants and plasticizers. They can be made from different raw materials. In the United States, BCAs are wood product derivatives, while in Japan they are made from petrochemicals. The Japanese BCA is much more expensive than the U.S. product. In each case, some unreacted starting material remains in the final product; this residue may degrade capacitor performance. Unfortu-

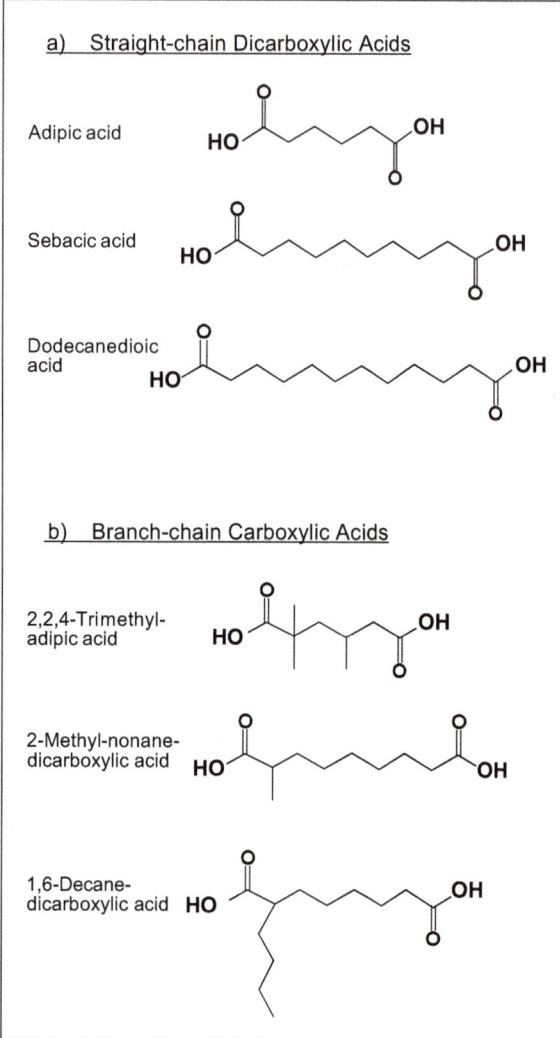

Fig. 2. Organic acid structures.

nately for U.S. capacitor manufacturers, the problem is much greater with our domestic BCA and limits use of those chemicals in capacitors.

Butyrolactone: Ethylene Glycol Electrolytes

We have developed electrolytes for use at high voltage and high temperature that use a mixed solvent with BL as the major component and EG as the minor component. With this solvent, both DCA and domestic BCA give good performance. The electrolytes have operated successfully in capacitors to 550 V, 105°C.

These electrolytes were developed under a program supported by NASA to make a hermetically sealed high voltage aluminum electrolytic capacitor [1]. The design of those capacitors and their performance are the subject of another paper at this conference [2]. The electrolytes are suitable also for commercial capacitors with conventional package design.

BL was chosen as the primary solvent component in order to get a wide operating temperature range. A mixed solvent is used because the resistivity of a BL electrolyte is markedly reduced by addition of a small amount of EG. This is because BL and EG have different solvating properties; BL is a basic solvent and EG is a protic solvent. A small amount of EG enhances solute dissociation and solubility. The best combination

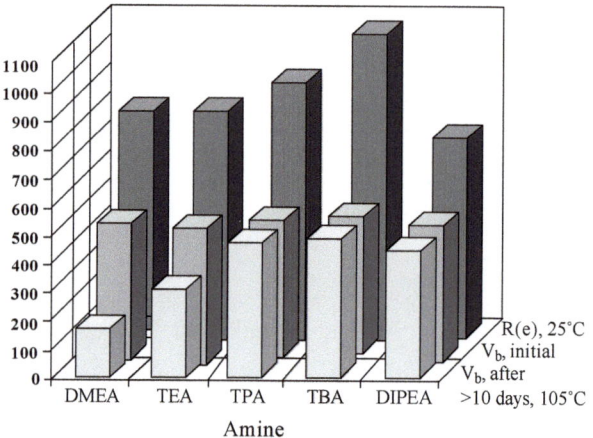

Fig. 3. Effect of amine on resistivity (R_e, ohm-cm) and breakdown voltage (V_b). Key: DMEA=dimethylethylamine; TEA=triethylamine; TPA=tripropylamine; TBA=tributylamine; DIPEA=N,N-diisopropylethylamine.

of low resistivity and thermal stability is found with a 9:1 BL:EG solvent.

The solute is a tertiary amine and a carboxylic acid. Selection of the amine is critical to good performance. Fig. 3 shows properties of five electrolytes, identical except for the choice of tertiary amine. All of them contain 0.45 M amine and 0.3 M sebacic acid in 9:1 BL:EG with 3% water. Initial resistivity depends on the size and structure of the amine. The initial breakdown voltage (V_b) is about the same for all compositions. Amines can react with BL to produce substituted butyric acids. These are relatively strong acids and reduce the voltage capability of the electrolyte. That reaction is the cause of the decrease in V_b after 10 days at 105°C, shown in Fig. 3. Voltage stability improves with increasing amine size. The best combination of low resistivity and stable V_b is obtained with DIPEA. A U.S. and foreign patent has been issued for electrolytes made with this amine [3].

Comparison of BL and EG Electrolytes

BL-based electrolytes are not well known for high

voltage applications. There are differences between BL and EG electrolytes that are important both to the capacitor designer and also to the capacitor user. These are discussed below.

Resistivity and ESR Increase

At elevated temperature DCA and BCA react with EG to make an ester, which is nonconductive. In both EG and BL:EG electrolytes, there is an esterification reaction between the organic acid and glycol. This increases electrolyte resistivity and is the primary cause of ESR increase during capacitor operation at high temperature. In EG electrolytes, this reaction continues as long as the capacitor is at elevated temperature, and the steady increase in ESR may limit operating life. In BL:EG electrolytes, the resistivity rises initially, then levels off. This behavior is compared for typical electrolytes in Fig. 4. The small increase in resistivity of the BL:EG electrolytes is responsible for the stable ESR observed during extended load tests.

For EG electrolytes, the rate of resistivity increase has a certain temperature dependence, as shown in Fig. 5. The temperature dependence is the same for an electrolyte made with a BCA (CP-42) as for an electrolyte made with DCA (BC-13B). A remarkable characteristic of

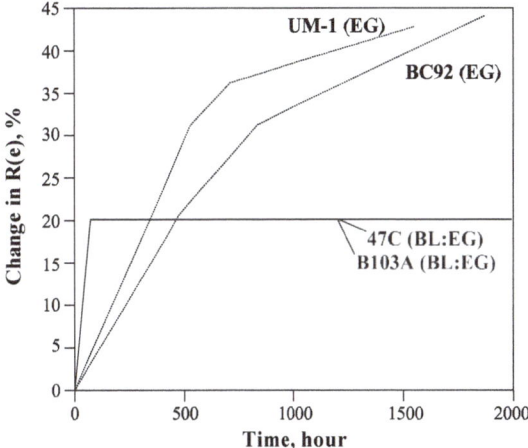

Fig. 4. Resistivity increase at 105°C. Key: UM-1= EG/BCA; BC92=EG/DCA; 47C and B103A= BL:EG/DCA.

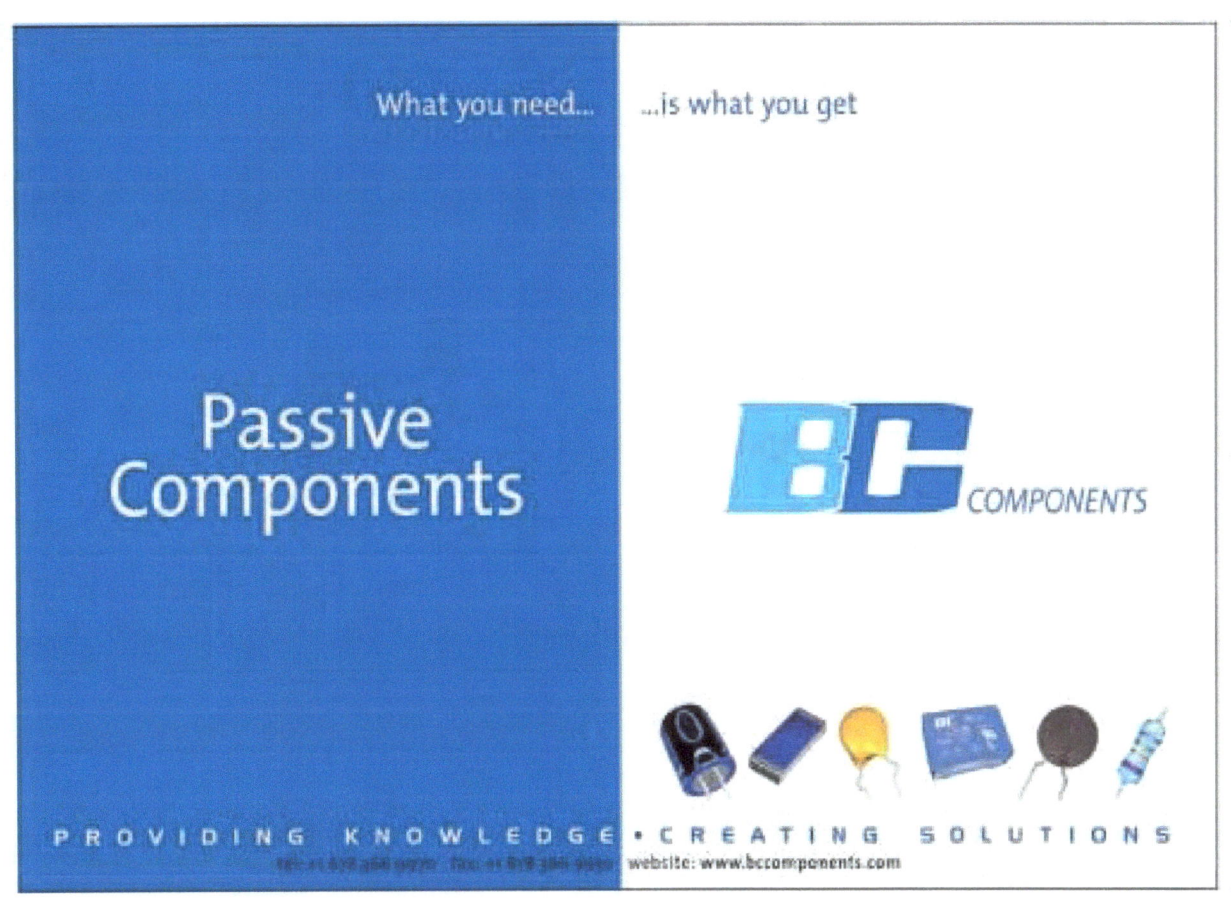

these BL:EG electrolytes is that the resistivity increase seems independent of temperature. In Fig. 5 we see that the resistivity increase for a typical formulation remains at 20% from 85°C to 125°C.

The BL:EG formulations exhibit much better resistivity stability than EG electrolytes. We expect this to result in more stable capacitor ESR and greater tolerance for temperature excursions during operation.

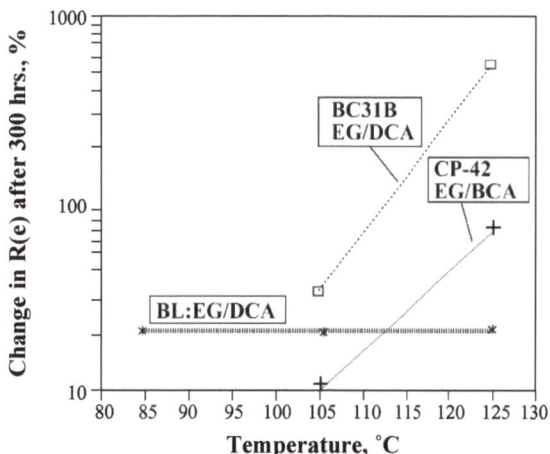

Fig. 5. Temperature dependence of resistivity increase.

Gas Generation

In EG electrolytes, the leakage current generates H_2 at the cathode in an amount proportional to the charge passed. In BL:EG electrolytes, a reduction reaction involving BL occurs which consumes some of the charge. This reduction product is soluble in the electrolyte so less gas is generated.

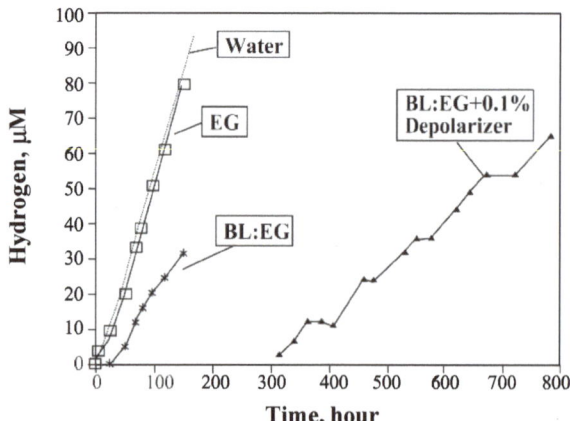

Fig. 6. Hydrogen gas generation at constant current of 30 μA.

The amounts of gas generated in EG and in BL:EG electrolytes is shown in Fig. 6. The data are for a laboratory experiment using wound sections in a glass cell that allowed measurement of very small gas volumes. The EG electrolyte generated the theoretical amount of H_2 for reduction of water, whereas in the BL:EG electrolyte only 40% of this gas volume was produced. When a nitroaromatic compound was present as depolarizer, no gas was produced until that compound had been consumed; then gas was generated at a reduced rate.

The inherent low gassing characteristic of BL is an attractive feature for capacitors exposed to severe conditions, such as high ripple current. For the hermetic capacitor, sufficient nitroaromatic was added to eliminate gas generation over the capacitor lifetime.

Chloride Sensitivity

Stringent requirements are placed on capacitor materials to make sure that chloride impurities are at very low levels. Trace amounts can lower the voltage capability of the electrolyte, possibly causing arcing and failure by shorting. It is desirable to have a robust electrolyte that can tolerate some ppm chloride without catastrophic effect. The BL:EG electrolytes are particularly good in this regard. Fig. 7 shows the chloride tolerance of several electrolytes, determined by measuring breakdown voltage after incremental additions of ppm chloride. The breakdown voltage of BL:EG electrolytes is relatively insensitive to chloride, whereas the commercial EG/BCA electrolytes show high sensitivity.

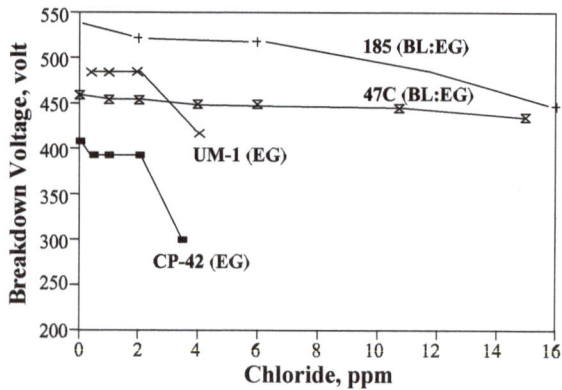

Fig. 7. Dependence of breakdown voltage on chloride concentration. Key: UM-1 and CP-42= EG/BCA; 47C and 185=BL:EG/DCA.

Toxicity

Comparing the three electrolyte solvents—EG, BL and DMF—only BL is not regulated by the Department of Transportation. EG is classified as toxic, and DMF is

Continued on page 22

Materials Research Furnaces, Inc.

Introduction:

Materials Research Furnaces, Inc. was founded in 1990, by a group of highly trained and experienced engineers and technical personnel from within the vacuum & high temperature furnace field. MRF Inc. was established to answer the challenge of the Research & Development community to produce the finest, high temperature, high vacuum and controlled atmosphere furnaces in the industry.

MRF BUILDING

MRF Inc. has supplied new furnace systems and replacement parts to Universities, National Laboratories and to private industries around the globe. A large part of MRF's Inc. services is providing parts and services for other furnace manufacturers' systems. MRF Inc. has over 14,200 square feet of space for manufacturing, assembly, engineering, sales and after market support.

Materials Research Furnaces, Inc. takes great care in providing you, the customer, a quality product that is both reliable, simple to operate and user friendly. The operation of any MRF Inc. system can be mastered in just a few hours. MRF Inc. furnace systems will complement any laboratory or manufacturing facility.

TANTALUM CAPACITOR SINTERING FURNACE

Products & Furnace Outline:

MRF Inc. produces a wide range of furnace for almost every application. Our furnaces range in temperature from 600 degrees Celsius to 3000 degrees Celsius. Our vacuum furnaces are designed to the 10-9 torr range. These furnaces can be used in a variety of inert gasses and other volatile gasses such as hydrogen and methane. Some of MRF Inc's furnaces are:

Continuous Belt Furnace: Top & Bottom Loading Furnaces:
Physical Testing: Front Loading Batch Sintering:
Hot Pressing 1 through 100 Ton: Graphite Tube Batch
Arc Melting Furnace: Crystal Growing Furnaces:
Muffle Tube Furnace: N2 BME Furnaces:

For more information:
Contact Materials Research Furnaces, Inc.,
Suncook Business Park; Rt. 28 & Lavoie Drive
Suncook, NH 03275: Attn: Daniel J. Leary, SM
Phone: (603) 485-2394 Fax: (603) 485-2395;
E-mail: mrf@interserv.com
WEB SITE: www.mrf-furnaces.com

Reps in Europe & overseas representation:
In the European Union contact:
Instron SFL
Severn Furnaces, Ltd.
Mr. Stephen Horrex
Brunel Way
Thornbury, Bristol BS 35 3UR, UK
Phone: 1454-414600; Fax: 1454-413277;
E-mail stephen_horrex@sfl.instron.com
All other overseas contact MRF directly.

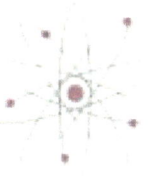

TANTALUM CAPACITOR SINTERING FURNACE

"TODAY'S FURNACES FOR TOMORROW'S TECHNOLOGIES"

Avnet Electronics Marketing Enhances Online Selector Tools

The World Wide Web's most comprehensive capacitor selection site, Capacitors Online ℠, has been recently improved for ease of use and flexibility. The tool has been created for purchasers and designers to locate the ideal capacitor for their application from a database of over 20,000 capacitors. There are two search methods. The first is by entering parameters like type of capacitor, capacitance value, voltage rating and tolerance. The second is a search by manufacturer part number or partial part number.

The tool also includes links directly to supplier data sheets, and price and availability are just a click away.

Enhancements include:

- Easier to use, flexible search parameters

- Side by side comparisons of similar components

- Online technical support

- Online ordering

- "Smart Results"
  If invalid parameters are selected, recommended selections will be offered as possible to the desired search parameters.

"Automation of passive part descriptions has enabled Motorola Supply Management to streamline the selection of industry standard items stocked by Avnet Electronics Marketing, reducing time-to-market," states David Saunders, Senior Buyer, Motorola Computer Group, Tempe, Arizona. "In addition, availability of detailed descriptions provides viewing of part numbers at a glance, allowing flexibility and accuracy to support both internal and external customers." When design engineers are selecting capacitors, there are many variations of the same device, making it difficult and a time consuming process. Typically, to check the price differences between components, the engineer compiled their list and submitted it to purchasing for pricing and availability. This information is now conveniently available on the Internet 24 hours a day, which saves time for both purchasing and engineering. Avnet Electronics Marketing has a wide breadth of capacitors from the following world leaders to choose from: AVX Corporation, Murata, Nichicon, Philips, Vishay Roderstein, Vishay Sprague, and Vishay Vitramon to name a few. If you cannot find the specific device you are looking for, click on the Technical Specialists button and our capacitor experts will respond within 24 hours.

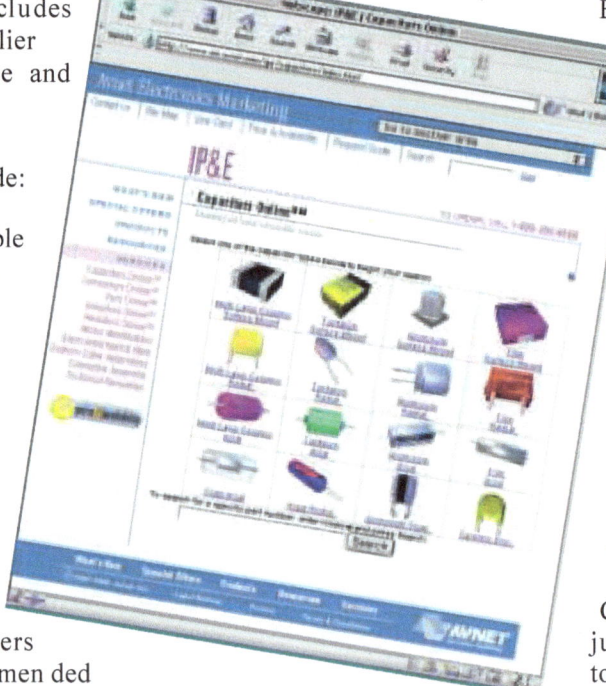

Capacitors Online℠ is just one of the selector tools on the Avnet Electronics Marketing IP&E Web site (www.ipe-tools.com) for interconnect, passive and electromechanical devices.

So try our new and enhanced Capacitors Online℠. Just one more reason that proves Avnet Electronics Marketing is committed to your passive component needs. As always, we would like to hear any feedback you may have. You may submit comments by clicking on the Contact Us button on any page within the site or call us at 1.888.IPE.PLUS.

W do
all the digging.

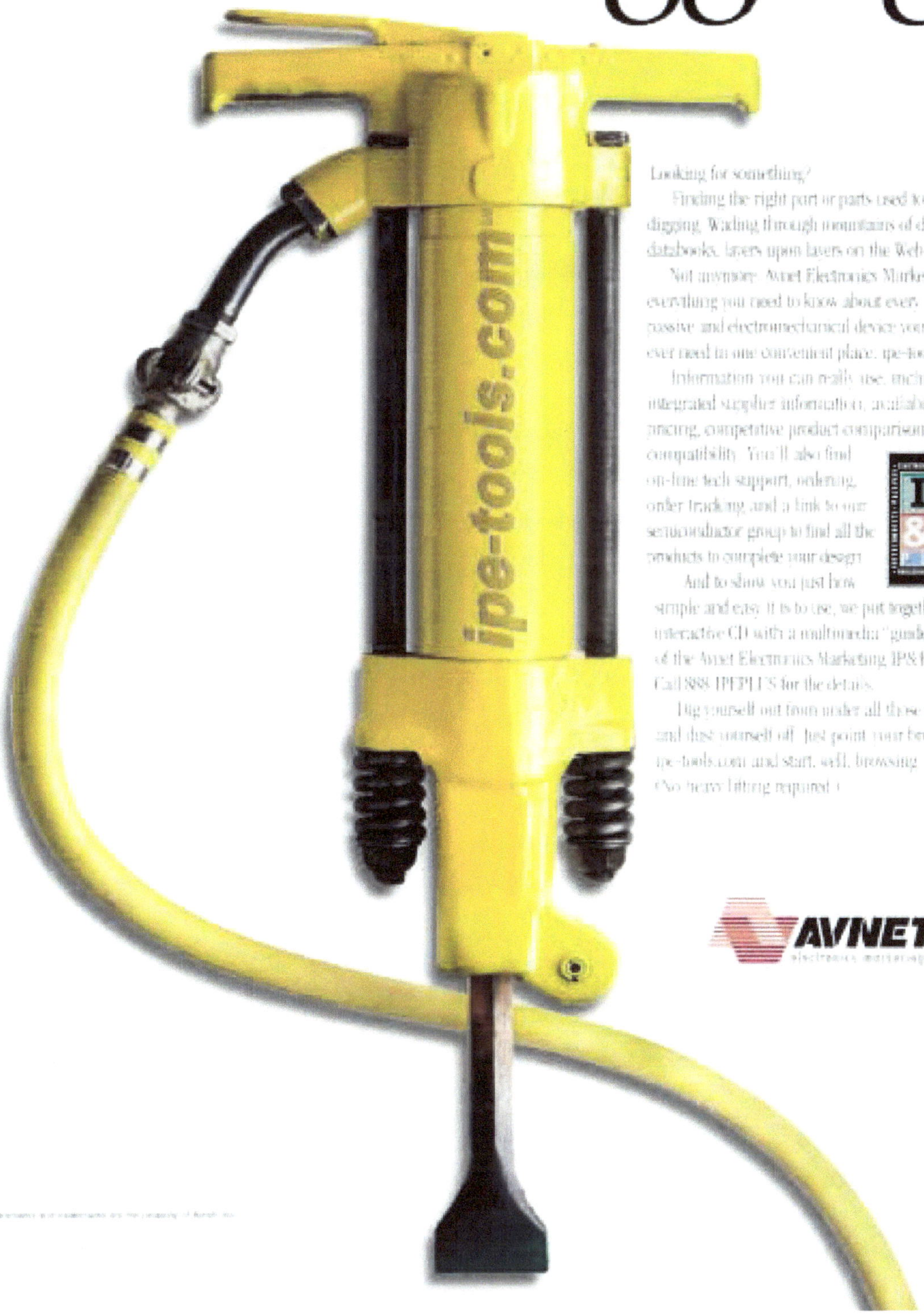

Looking for something?

Finding the right part or parts used to take some digging. Wading through mountains of datasheets, databooks, layers upon layers on the Web.

Not anymore. Avnet Electronics Marketing has put everything you need to know about every interconnect, passive and electromechanical device you'll probably ever need in one convenient place: ipe-tools.com.

Information you can really use, including integrated supplier information, availability, specs, pricing, competitive product comparisons and compatibility. You'll also find on-line tech support, ordering, order tracking, and a link to our semiconductor group to find all the products to complete your design.

And to show you just how simple and easy it is to use, we put together a free interactive CD with a multimedia "guided" tour of the Avnet Electronics Marketing IP&E sites. Call 888-IPEPLUS for the details.

Dig yourself out from under all those datasheets and dust yourself off. Just point your browser to ipe-tools.com and start, well, browsing. (No heavy lifting required.)

AVNET
electronics marketing

Continued from page 18
suspected to be carcinogenic and teratogenic.

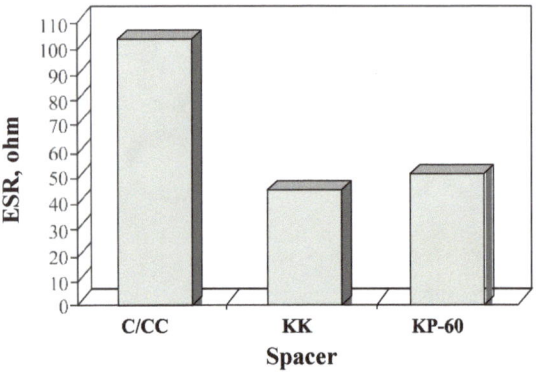

Fig. 8. Resistance of electrolyte-saturated spacer coupon with BL:EG electrolyte. Key: C/CC=calendared Kraft; KK=uncalendared Kraft; KP-60= Kraft/polypropylene blend.

Spacer Wetting

Glycol bonds strongly to cellulose fibers. This promotes thorough wetting that reduces the spacer resistance, even of highly calendared paper. Dense paper is not easily wet by BL, and the small amount of EG in the BL:EG electrolytes does not help significantly in this regard. This results in a high spacer ESR, even with a low resistivity electrolyte. With low density spacer, such as uncalendared Kraft, there is no problem with ESR, but this cannot be used as the sole spacer at high voltage. A spacer consisting of a blend of polypropylene and Kraft fibers (KP series from Mt. Holly Dielectric) provides low ESR and good voltage support. The synthetic blend has been used in 600 WV ratings with no shorts at aging or burn-in. The resistances of these different spacers with a BL:EG electrolyte are compared in Fig. 8.

Cover Compatibility

Ethylene glycol is inert toward all usual polymer cover materials. In contrast, BL is a relatively strong solvent for some rigid cover materials. It swells phenolic

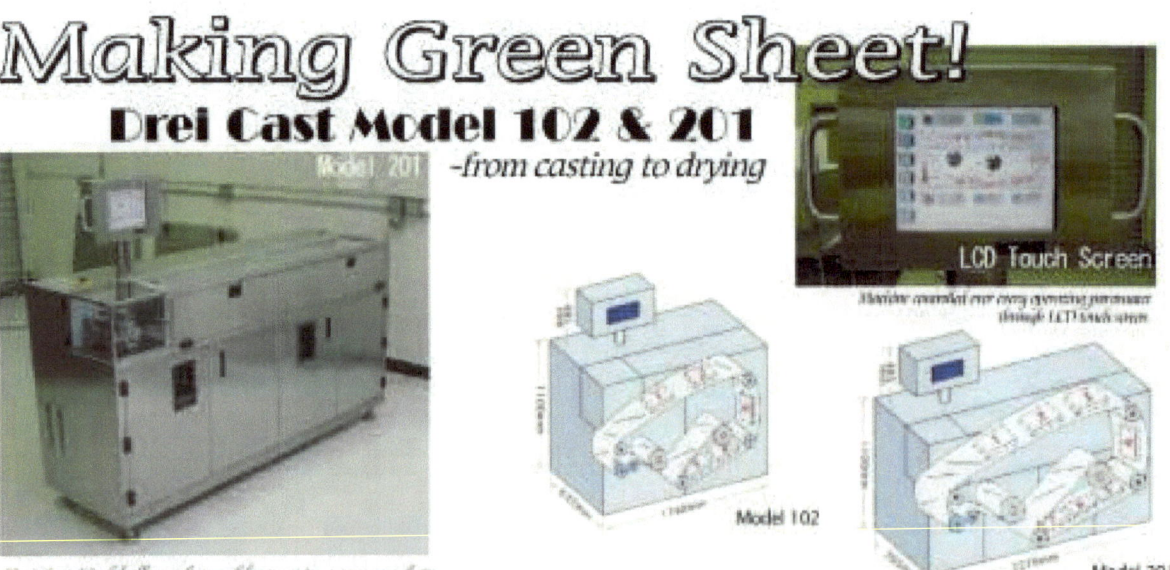

and may leach chloride and other trace impurities. The mechanical strength and rigidity of nylon covers is not compromised by BL, but trace impurities present in a nylon used in certain low voltage covers were found to cause problems at high voltage. With BL electrolytes, careful selection of cover material is critical. But BL is less aggressive than DMF, and that solvent has been used in electrolytic capacitors for years.

Capacitor Test Results

The result of our development effort is the family of electrolytes shown in Table 1. They provide wide temperature range and low resistivity (ρ) with high breakdown voltage (V_b). These electrolytes have been extensively tested in capacitors at 105°C. Capacitors were designed for voltage ratings of 250, 450 and 540 V. This last rating was for a NASA application at 270 V at 50% derating; the same design will work equally well at 550 V. Life test performance of capacitors with nonhermetic package design is shown in Fig. 9. The 250 V rating is in a CDE Type MLP flatpack design. The 450 and 540 V ratings are conventional large can packages with nylon covers and screw terminals. Performance in the hermetic package is described elsewhere [2].

Table 1:
High Voltage BL: EG Electrolytes

	WV	Temp (°C)	ρ, 30°C (Ω–cm)	V_b
47C	350 V	–55–+105	575	460 V
B090A*	350 V	–55–+105	570	460 V
179	450 V	–45–+105	700	520 V
BC158B	500 V	–45–+105	880	550 V
B103AD	550 V	–45–+105	1070	600 V

*Can be used with dense spacer.

The electrical parameters of the 250 WV rating were essentially unchanged during 2,000 hours at 105°C. The DF of these capacitors remained at 4.5%. The initial DF of the 450 V rating is 4.4%. Over the 4,000-hour test, the ESR ranged from a minimum of 195 mohm to a maximum of 265 mohm. The ESR and LC of the 540 V capacitors decreased early in the test. We think the initial values of this particular test group are high due to insufficient aging, because capacitors made for other tests had initial values about the same as shown here for 250 hours. At 250 hours the DF is 7.2%. Even including the initial high ESR, the overall range in ESR

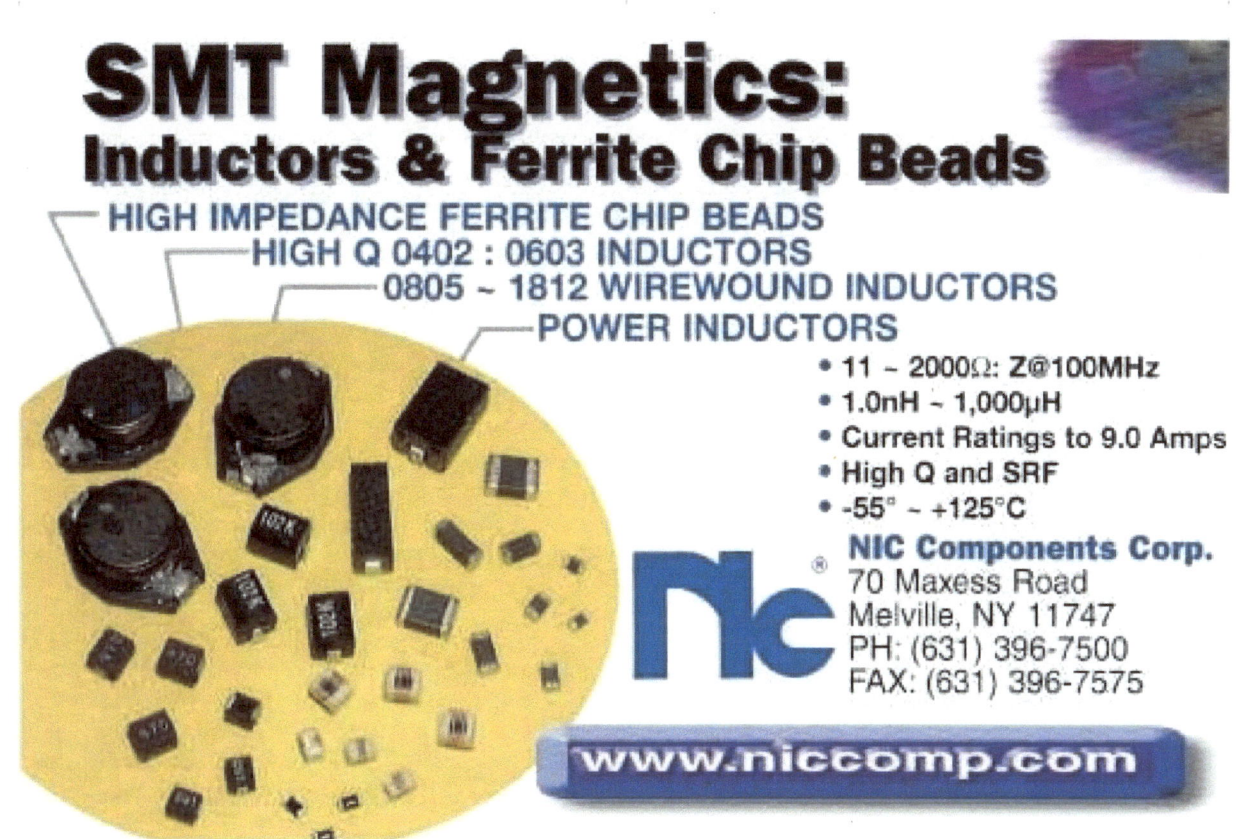

over 4,000 hours was from 300 to 400 mohm.

A 600 V electrolyte was tested for 2,000 hours in capacitors with the same design as the 540 V rating, but run at 600 V, 85°C. There were no shorts at aging, and the initial DF was 9.6%. Work is in progress to reduce DF and leakage current. At this voltage, the capacitor performance may be limited by anode foil properties as well as by electrolyte. This clearly is the frontier to be conquered.

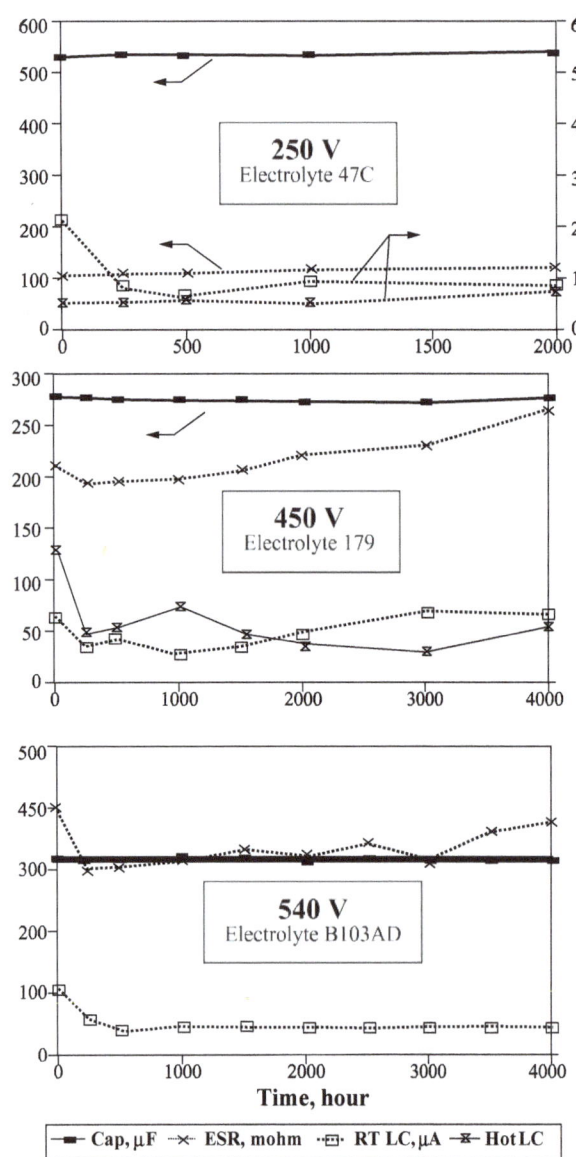

Fig. 9. Capacitor performance on load tests at 105°C. Hot LC monitored during test.

Conclusions

High voltage capacitors containing BL:EG electrolytes performed very well in capacitor tests at 105°C, exhibiting excellent parametric stability. Comparison with typical EG electrolytes indicates that in severe applications, these new electrolytes will have better ESR stability, less tendency for gas generation, and better ability to withstand high temperature and trace chloride. They seem particularly well suited for motor drive applications and are available for capacitor ratings up to 550 WV.

Acknowledgments

This work was carried out for NASA Marshall Space Flight Center under contracts NAS 8-38899 and NAS 8-39371, awarded under the SVIR program. Mr. William Elias of Bace Technologies did the capacitor design and engineering and also supervised the capacitor test program. Capacitor assembly and testing were done at Cornell Dubilier Electronics.

References

1. Final Report, "Hermetically Sealed Aluminum Electrolytic Capacitor," Contract NAS 8-39371, NASA Marshall Space Flight Center, September 1995.
2. L.L. Macomber and W. Elias, "Longlife, High Voltage Hermetically Sealed Aluminum Electrolytic Capacitors," CARTS 96, March 1996.
3. Y. Liu, U.S. Patent 5,496,481, March 5, 1996.

Publisher

Continued from page 5

ply situations), then customers who sign long term agreements today may not be competitive when the capacitor industry downturn occurs in 2001.

Such long term agreements make tremendous economic sense for the capacitor manufacturers because they work to keep price erosion at a realistic level. Therefore, capacitor manufacturers can generate enough profits to reinvest in incremental capacity expansion in slow years. The fear of the capacitor manufacturers, however, is that when the capacitor industry takes a turn for the worst, OEMs and CEMs who had entered into long term agreements will abandon their set rates of price erosion in order to remain competitive. In short, a capacitor manufacturer who sells $500 million or $1 billion worth of capacitors in one year cannot afford to lose an account to a company that sells more product than the value of the entire passive electronic component industry.

— *Dennis M. Zogbi*
Publisher, *Passive Component Industry*
President, Paumanok Publications, Inc.

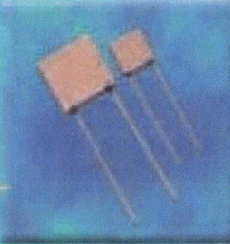

FEATURE

Anode and Cathode Foil: Market Dynamics of Thin Foil, Etching and Winding

The primary raw material consumed in the manufacture of aluminum electrolytic capacitors is thin aluminum foil. This foil must be etched and electrochemically formed before it can be wound as the dielectric support material for the finished aluminum electrolytic capacitor. Etching is the most expensive process in the production of aluminum electrolytic capacitors. Most major world manufacturers of aluminum electrolytic capacitors attempt to control their processing costs by owning and operating their own foil processing plants. But no major manufacturers of aluminum electrolytic capacitors own aluminum foil feedstock plants, so the general starting point in the supply chain is the purchase of the thin, high-purity foil direct from the merchant market.

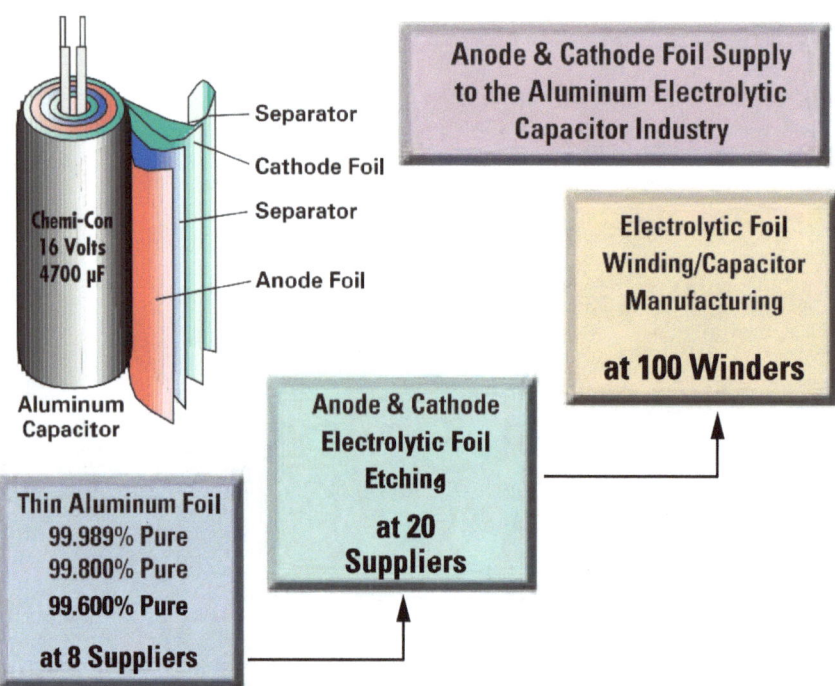

Foil Supply to the Aluminum Electrolytic Capacitor Industry

Generally, there are three valued-added steps in supplying aluminum foil to the global capacitor industry. These include the supply of thin, high-purity aluminum foil; etching of that foil into either anode or cathode foil; and winding of that foil into capacitors.

Thin foil suppliers provide solid-core and porous foils to the etchers, who expose the foil to an electrochemical process in which the metal is dissolved to increase the surface area of the foil by creating a dense, interconnecting network of small channels. This process involves running the thin aluminum foil stock through a chloride solution with an AC, DC or AC/DC voltage applied to the etch solution and the aluminum foil. The increase in the surface area of the foil is known as "foil gain." Low voltage foil gains can be as high as 100 times, and high voltage foil gains can be as much as 25 times the standard capacitance value of the beginning thin foil.

Thin Aluminum Foil (Porous and Solid-Core)

The requirements for thin foil call for 99.96% purity, although 99.8% and 99.9% purity foils are typical. Very low leakage current capacitors require 99.989% purity foils.

Thin aluminum foil costs approximately $2.00 to $3.00 per pound (not square meter) from Asian, European and American merchant suppliers. The difference in price depends upon the foil's thickness, weight and purity level. Capacitor etchers generally purchase gauge thicknesses between 0.0015 and 0.010 inches, depending on the voltage rating. These foils are sold in rolls that are usually 19 or 20.75 inches wide, although widths as high as 40 inches are also available.

Thin Aluminum Foil Suppliers

Thin foil suppliers in Japan include Toyo Aluminum, Showa Aluminum, Sumitomo Light Metal and Nippon Light Metal. Toyo Aluminum is the most active thin foil supplier, catering to the aluminum electrolytic capacitor industry in Japan; the company reportedly supplies the largest quantity of porous and solid-core foils for consumption by foil-etching houses.

In Europe, thin foil suppliers to the aluminum electrolytic industry include V.A.W. in Germany, Pechiney in France and Lawson Mardon Singen in Germany. V.A.W. is the largest thin foil stock supplier for European consumption.

Continued on page 30

EVENTS

Ferro Celebrates New Vista Facility with Impressive Grand Opening Event

Ferro Corporation, one of the world's leading providers of specialty industrial and electronic materials, celebrated the recent opening of its new, state-of-the-art electronic materials facility in Vista, California, with open house festivities June 29 and 30. The event was attended by customers and colleagues from across the country as well as industry media and local dignitaries. Activities on Thursday included a golf tournament and an informal tour of Coronado Island, winding up with a cocktail party and gala dinner reception Thursday evening. Friday began with the ribbon-cutting ceremony and concluded with lunch.

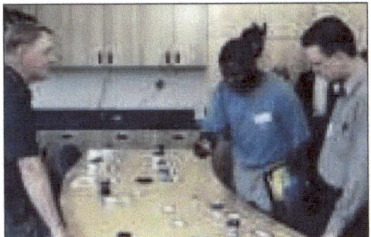

John Ekis, Ferro Product Manager [left], discusses innovative materials samples with Maurice Watson [center] and Michael Walker in the Vista facility's main conference room.

Gene Thomas, Ferro Asian Sales Representative [front left], shares insights with Ferro customers.

The main emphasis of the entire event was a highly informative guided tour of the new facility. The Vista

Ferro consultant Ann Paine and her husband Bill [left and center] enjoying themselves during Thursday evening's dinner reception.

plant devotes nearly 100,000 square feet of space and the synergistic efforts of over 100 employees to the R&D, manufacturing and testing of innovative materials and technical processes for the electronics industry. The tour covered more than a dozen key areas of interest. Each department was hosted by knowledgeable Ferro professionals who provided a general overview of their department and answered questions from attendees. ❑

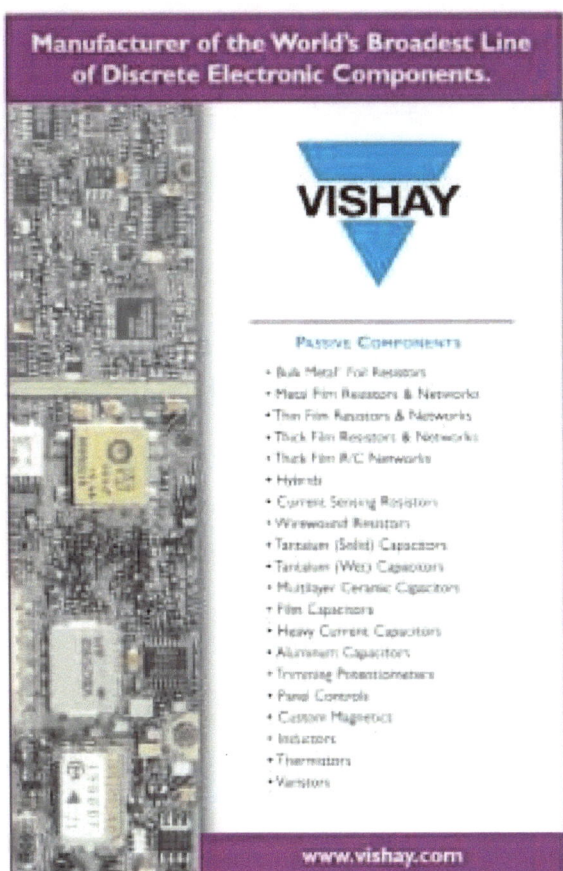

Events

Craig Benson, FEM World Wide Business Manager [left], Mark Skoog, Business Unit Manager [center], and Robert Rieger, Vice President, Colors, Coatings and Electronics, participating in the ceremonial ribbon-cutting on Friday morning.

Robert Rieger, Vice President, Colors, Coatings and Electronics, addressing guests during the ribbon-cutting ceremony.

Ferro Human Resources Manager Keith Michaelis [left] responds to an observation by colleague Jim Henry, one of Vista's LTCC engineers.

An assemblage of attendees posing for posterity before the tour begins.

A partial view of the Vista plant's manufacturing area.

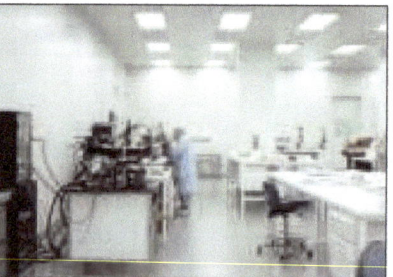

Ferro's Class 10,000 MLCC/LTCC Clean Room supports application engineering, development and quality control processes.

Head 'em up! Move 'em out!

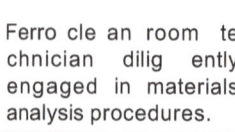

Ferro clean room technician diligently engaged in materials analysis procedures.

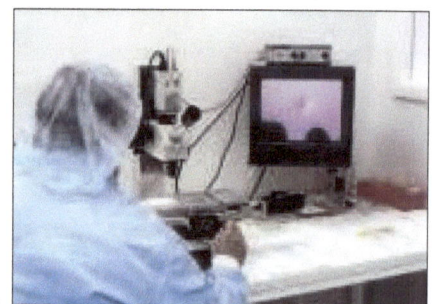

28 PASSIVE COMPONENT INDUSTRY JULY/AUGUST 2000

Events

ECA Summer Conference

Dennis M. Zogbi, president of Paumanok Publications, Inc., discusses markets, technologies and opportunities for passive components at ECA's Summer Conference 2000.

Dave McCurdy, EIA President, and Joe Matthews from Delphi discuss passive component content in automotive electronic subassemblies.

Bob Gourdeau, VP, Sales and Marketing, out spreading the word on BC Components.

Mike O'Neil (left) from Heraeus, trying to convince Mark Messow from Celestica to buy paste instead of MLCCs.

David Fancher (Vishay), John Rector (IBM), Martin Kris (Shoei) and John Ekis (Ferro). Pick a town, any town, and Martin will know the best restaurants.

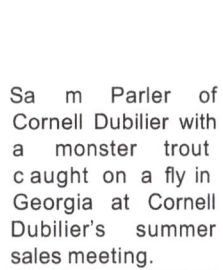

Mike Thompson of Taiyo Yuden says the way to go is high capacitance.

Sam Parler of Cornell Dubilier with a monster trout caught on a fly in Georgia at Cornell Dubilier's summer sales meeting.

Raw Materials

Continued from page 26

In the Americas, Ormet Aluminum Mill Products (Jackson, Tenn.), Alcoa Specialty Products Division (Badin, S.C.), and J.W. Aluminum (Charleston, S.C.) sell thin aluminum foil stock to foil etchers.

Etching of Thin Aluminum Foil

Thin foils must be electrochemically etched before use in fixed capacitors. Etching is expensive and costs approximately $7 to $13 per pound. Since the etching process is the most expensive in the manufacture of aluminum electrolytic capacitors, most of the large capacitor manufacturers attempt to control these costs by etching their own foils in-house. Still, a large merchant market for etched foil exists to supply the smaller aluminum electrolytic capacitor manufacturers and to "fill in the blanks" at the larger capacitor houses.

As is the case with most other fixed capacitor dielectric materials, aluminum electrolytic capacitor manufacturers will also look toward merchant market suppliers for better technology in etched aluminum foil.

In the industry, companies will refer to porous versus solid-core foils. Porous foils are etched much deeper than solid-core foils, generally etched to halfway through the foil (50% of thickness), as opposed to solid-core foils, which are etched "to less than halfway" through the thickness of the foil. The deep etching process used in porous foils enables the electrolytes to diffuse rather quickly from one side of the foil to the other. Porous foils offer less stable mechanical characteristics than solid-core foils and therefore are used only in applications in which high voltage, high capacitance and lower equivalent series resistance (ESR) is required in the finished capacitor. Thus, porous foils are usually found in photoflash, strobe and motor start aluminum electrolytic capacitors.

Solid-core foil accounts for the majority of foil usage in aluminum electrolytic capacitors because most aluminum electrolytics used in consumer electronics are 50 volts or less (usually 16 volts) and do not require lower ESR or less heating during charge and discharge. Solid-core foils

Japan/Asia	Europe	Americas
Toyo Aluminum (Japan)	V.A.W. (Germany)	Ormet (USA)
Showa Aluminum Alloy (Japan)	Pechiney (France)*	Norandal (USA)
Nippon Light Metal (Japan)		

*Pechiney is vertically integrated and owns the etching service Satma in France.

also have much better mechanical strength than porous etched foils.

Aluminum Foil Etchers

The world's top merchant suppliers of etched anode and cathode foils for the aluminum electrolytic capacitor industry are KDK in Japan and Becromal SpA in Italy. Both excel in supplying standard and specialty anode and cathode foils. There are many secondary suppliers who are very successful in certain market niches. Satma in France, for example, excels at making

Continued on page 32

EFC BRINGS LEAD TIMES BACK DOWN TO EARTH!!

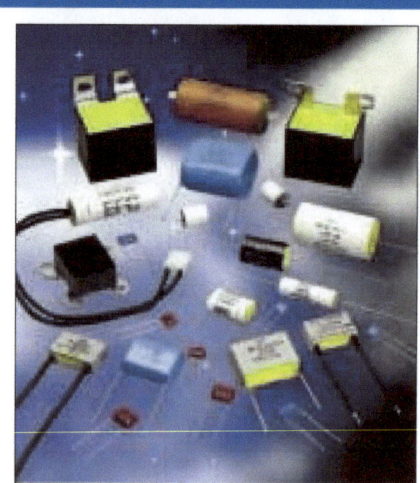

THE MISSION: TO BRING INTERPLANETARY LEAD-TIMES DOWN TO EARTH!

EFC, A LONG TIME LEADER IN FILM CAPACITOR TECHNOLOGY, HAS TAKEN ON THE MISSION AND IS OFFERING 5 WEEK LEAD TIMES ON MOST PRODUCTS.

WE SUPPLY POLYESTER, POLYPROPYLENE, POLYCARBONATE, AND EVEN POLYSTYRENE DIELECTRICS IN AXIAL AND RADIAL LEAD CONFIGURATIONS.

CONTACT US DIRECTLY FOR LITERATURE OR VISIT OUR WEBSITE.

ELECTRONIC FILM CAPACITORS — EFC

41 Interstate Lane • Waterbury, CT 06705 USA •
Phone: (203) 755-5629 • Fax: (203) 755-0659
E-Mail: efc@filmcapacitors.com • Website: www.filmcapacitors.com

Face Up to Your Capacitor Delivery Problems.

Get Standard and Custom Integrated Passive Devices from California Micro Devices in Less than 8 Weeks!

Avoid the capacitor shortage and slow deliveries by designing in CAMD's cost effective Integrated Passive Devices (IPDs). Instead of waiting endlessly for your discrete capacitors, you can count on our IPDs in four to eight weeks.

Along with capacitors, our IPD technology delivers other elements such as resistors, diodes and active circuitry in single chip solutions that enhance the performance of your entire system. You get the benefit of reduced component size for greater miniaturization in high-density printed circuit boards.

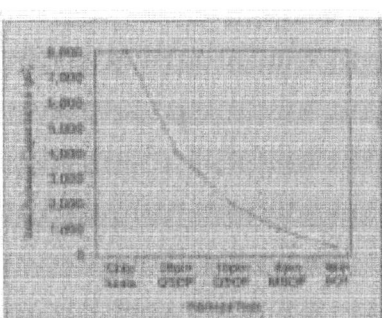

CAPACITANCE VALUE LIMITS

Additionally, the use of IPDs results in greater reliability and lower total cost.

For today's advanced technology requirements, CAMD will assist you in designing-in a standard or custom IPD solution. End the capacitor roadblock with California Micro Devices.

Call the Applications Dept. at 1-800-325-4966 or visit our web site at www.calmicro.com.

 California Micro Devices 215 Topaz Street, Milpitas, CA 95035 • Tel: (408) 263-3214
Fax: (408) 263-7846 • Email: cmd@calmicro.com • www.calmicro.com

MARKET STATISTICS

Changing Markets in Passive Component Distribution: 1999-2005

According to primary manufacturers of passive components, there is a substantial trend away from directly supplying original equipment manufacturers, and a move toward direct supply of contract electronic manufacturers and electronic component distributors. According to one major manufacturer of ceramic and tantalum capacitors, the company noted that within the first six months of 2000, their top 10 customers became entirely CEMs and distributor-based; no OEMs are included in the top 10. This marked a substantial change over the first six months of 1999 when five OEMs made up half of the top 10 customer base.

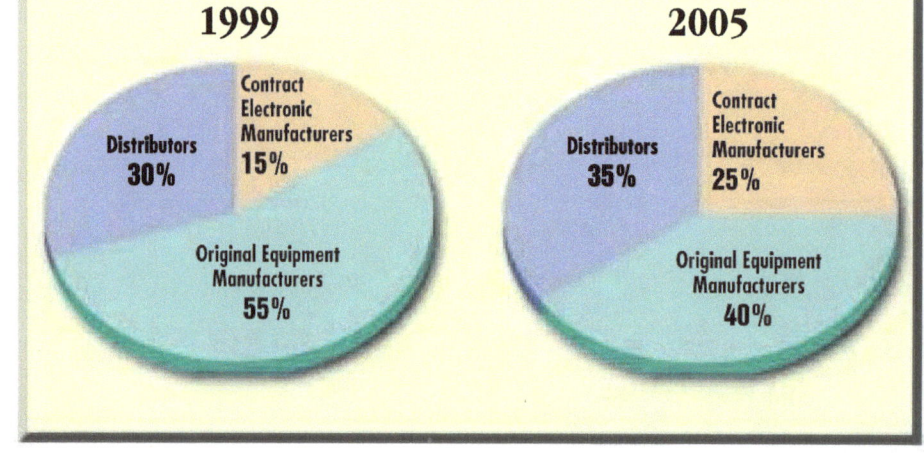

Because of this trend, it is believed that within five years shipments direct to OEMs will account for 40% of com ponent sales, down from 55% in 1999. Component sales will shift away from OEMs to CEMs, and distribution will continue to obtain more business because of value-added services.

OEMs will still exert tremendous influence over the buying process as they leverage their brandnames in order to get better pricing for their selected CEMs. Companies that should fare well during this transition include SCI, Celestica, Solectron, Jaybil and Flextronics; com ponent distributors who will fare well during this transition include TTI, Avnet, Future and Arrow. ❑

RawMaterials

Continued from page 30

motor start and strobe foils, while Japan Capacitor Company (JCC) in Japan is known for its photoflash foils, especially for camera and medical implant applications. Captive sources of etched foil include Matsushita, Nippon Chemi-Con (KDK), Rubycon and BC Components (Zwolle).

Raw Material Costs

Raw material cos ts for the production of aluminum electrolytic capacitors are generally less than those incurred in the manufacture of other dielectrics. The most expensive processes are associated with the etching and anodization of thin aluminum foil. Costs are high for etching because i t is technologically sophis- t icated; costs are high for anodization because i t is electricity-intensive.

Costs for other dielectrics are more expensive. Ceramic capacitors, for example, con sume palladium, silver and titanate materials, which are very expensive. (This explains why MLCC manufacturers are turning toward base metals such as nickel and coppe r for their electrodes.) Tantalum capacitors con sume tantalum ore, tantalum metal powder and tantalum wire, which are also very expensive materials. DC film capacitors consume metallized polyester and polypropylene and are very expensive in their thin gauge forms.

A com parison of raw material pricing shows that anode foils cos t about $10.00 per square meter and cathode foils cos t about $3.00 per square meter; tantalum powder cos ts about $180.00 per pound and tantalum wire cos ts about $210.00 per pound. Ceramic dielectric material averages about $10.00 per pound; palladium cos ts from $600.00 per troy ounce (depending on many factors). Polyester extruded to 4 microns will cos t about $130.00 per pound. Thus, aluminum foils are actually very cost-effective compared with other dielectric materials. This results in a low average unit price per microfara d and therefore a lower overall com ponent cos t in aluminum electrolytic capacitors. ❑

NEWSMAKERS

AVX Transient Voltage Suppressors: Ideal Alternative for Scarce Diodes

AVX Corp.'s TransGuard ® Series of bidirectional transient voltage suppressors (0805 size or smaller) are suitable substitutes for hard-to-fin d TVS diodes. Offering electronic circuit design engineers a reliable transient voltage protection solution, the TransGuard ® chips are ready for delivery today.

The smaller size TransGuard ® chips (0603 and the newly released 0402 sizes) are increasingly being specified by designers faced with densely-packed circuit boards in next-generation products, such as cellular phones, pagers and digital cameras. With the current shortage of diodes, our readily available 0603 and 0402 varistor chips offer a perfect alternative for transient suppression, providing design engineers with superior TVS characteristics over silicon alternatives, such as faster response time and multiple strike capability.

The miniature TransGuard ® Series 0402 chips are available in 5.6, 9, 14 and 18 volts, with a 50 millijoule energy rating, and a typical capacitance range from less than 90 p F at 18 V to 360 p F at 5.6 V. A StaticGuard ® low capacitance version offer s typical capacitance of 40 p F, with an energy rating of 20 millijoules. The 0402 chips are rated for operation over the full temperature range from -55°C to +125°C and are packaged in seven-inch reels com patible with high-speed placement equipment.

The 0402 TransGuard ® Series chips are packaged in seven-inch reels in 4,000 and 10,000-piece quantities and are compatible with high-speed placement equipment.

Typical pricing for the 0402 TransGuard ® Series chips ranges from $.122 to $.129 in quantities of 100,000 with a lead time of eight to 10 weeks ARO. For more information about AVX TransGuard ® products, contact AVX Sales and Marketing literature depart- ment at 843-946-0414, by fax at 843-448-1943, or at www.avxcorp.com on the Web.

Ne w Thin-Film Resistor Net w orks from Vishay Feature Compact, Surface Mount Package, ±0.025 Typical Ratio Tolerances

A new series of thin-film resistor networks offering absolute tolerances to ±0.1% and resistance ratio matching to ±0.025% have been released by V ishay Intertechnology, Inc. The new ORN series dividers are

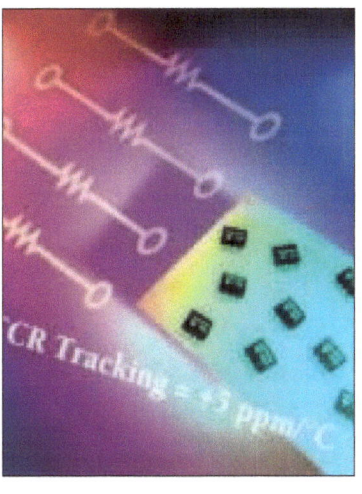

packaged in a JEDEC standard MS-012 surface mount package measuring 0.157 inch (3.99 mm) by 0.195 inch (4.93 mm), with a 0.068 inch (1.73 mm) height profile, making them a compact alternative to discrete implementations in circuits using a differential op-amp.

Built on Vishay's thin film technology, devices in the ORN series integrate two pairs of resistive elements with standard R1/R2 divider ratios of 2, 5, 10, 20, 25, 50 or 100. The typical temperature coefficient is ±25 ppm/C with a tracking temperature coefficient ratio of just ±5 ppm/C.

ORN resistor networks easily substitute for discrete com ponents by matching an integrated version of the

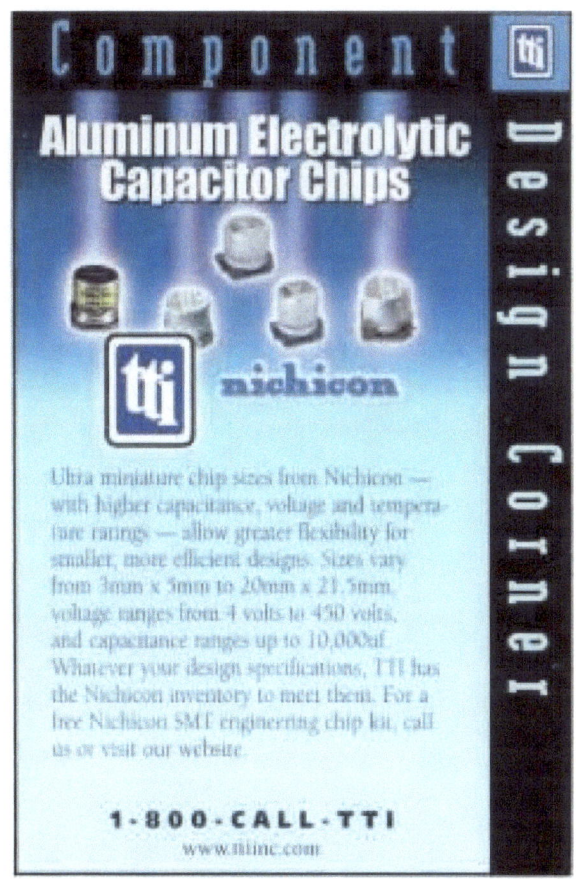

existing resistor schematic to the discrete implementation. Vishay streamlines this process by providing a wide choice of off-the-shelf standard schematics and resistance values from 1 kilohm to 100 kilohms

"The new ORN series combines optimal ratio precision, small size, and exceptional stability for most applications," said Bill Brennan, Vice President of Marketing and Sales for Vishay's High Precision Products Group. "In space-constrained designs, the ORN series will lower the installation cost per element by 75% compared to discrete resistors and provide a higher level of reliability."

Samples and production quantities of the ORN series are available for immediate delivery, with lead times of up to four weeks for larger orders. Contact: Andrew Post, Communications Manager; phone: 610-251-5287; fax: 610-889-9429; email: andrew.post@vishay.com.

New 550 VDC Large Can Electrolytic Capacitor

Evox Rifa has just extended the voltage range of their PEH200 Series Large Can Electrolytic Capacitors to 550 VDC. With a transient voltage allowance of 800

volts and a very high ripple current capability, the capacitor is ideally suited for power electronics applications such as in power generators, UPS, welders and motor drives. PEH 200 is uniquely designed for low ESR and excellent thermal transfer, giving very long life at elevated temperatures. The PEH 200 series is now available in voltages from 25 to 550 VDC with capacitance values up to 330,000 mF.

Cost: About $25 each, depending on model and quantity. Delivery: Stock to 10 weeks. Contact: service@evox-rifa.com.

New Ultra-Miniature Resistor is in Tiny 0201 Package

Kamaya, the thick-film technology leader, has introduced a new line of 1/20 watt resistors in a 0201 package (0.6 mm x 0.3 mm). The RMC1/20 series resistors are ideal for high density applications. They require only 40% of board space, when compared to a 0402 package. Applications include oscillators, notebook PCs, PC cards, digital cameras and small hand-held devices such as pagers, cellular phones and palm tops.

The RMC1/20 has a maximum working voltage of 25 V. They are available in F (±1%), G (±2%), or J (±5%) tolerances. The resistors are packaged on an 8-mm press-pocket carrier, with a dimensional accuracy of ±0.03 mm. This helps to improve the integrity of the placement of the resistor. According to Mike Liebing, Manager of Marketing and Sales, North America, "The explosive growth of the small, hand-held, battery-powered products has driven the market demand for tighter PCB packaging. This is what spurred the development of the RMC1/20."

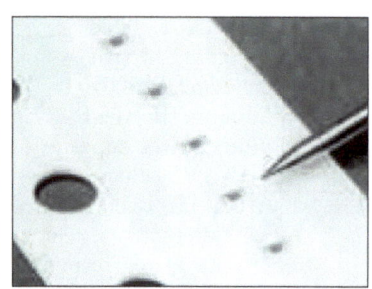

Cost for 100,000 pieces of a RMC1/20101JPA tape and reel is $30.00/thousand. Delivery is stock to eight weeks. Data sheets and free engineering evaluation samples are available. For further information call Kamaya at 219-489-1533 or visit their Web site at www.kamaya.com.

SMD Air Wound Coils for High Power RF Applications

The 291 through 294 Series of SMD Air/Wound Coils from Frontier Electronics are designed for high power RF applications such as pagers, cell phones and mobile radios.
Key Specifications:

Inductance values: From 2.4 nH to 538 nH.
Q-values: Over 100.
SRF: Greater than 3 Ghz.

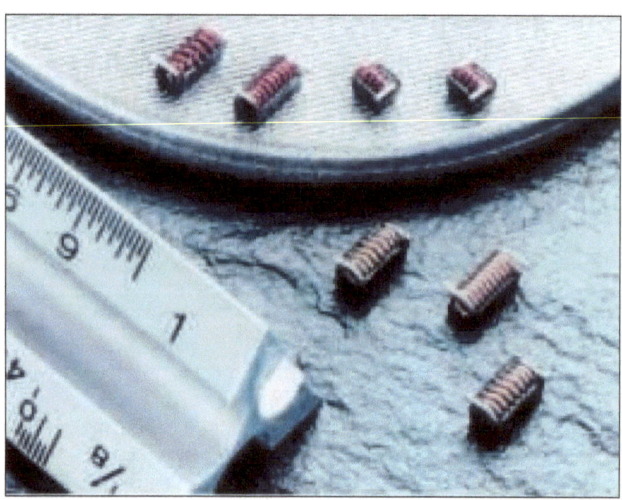

Newsmakers

Six package sizes are available. Tape and reel packaging with a flat top allows for pick-and-place mounting.

Price: Starts at under $.25 each at OEM levels, depending on quantities. Delivery: 6-8 weeks.

More information: Marketing Department, Frontier Electronics, 685 E. Cochran St., Simi Valley, CA, 93065; (800) 929-9888; fax (805) 522-9989; www.frontierusa.com.

Laube Technology Ceramic Chips are ISO9002 Certified

Key Features and Specifications :
 ISO 9002 certified.
 Capacitance ranges from 0.10 pf to 4.7 µf.
 Working voltage up to 100V.
 Tolerances from 0.10 pf to +100%/-0%.
 Operating temperature range -55°C to +125°C
 Standard termination is nickel barrier, silver optional.

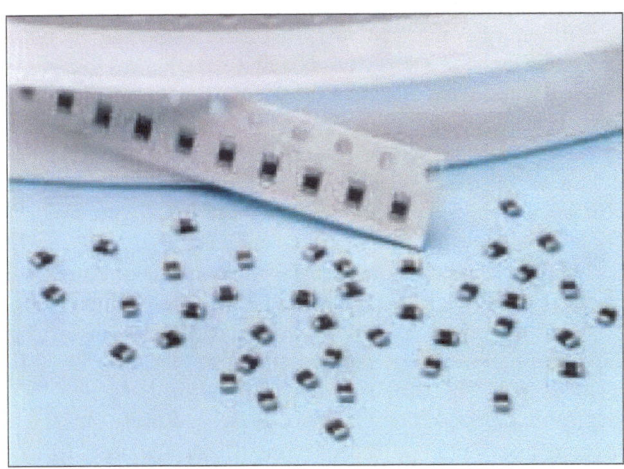

Will withstand wave, reflow and vapor phase soldering.

Price: Starts at $0.0075 each in OEM quantities, depending on size, capacitance and tolerance.

Delivery: Stock to eight weeks.

For more information: Laube Technology, 550 Via Alondra, Camarillo, CA, 93012; (888) 355-2823; fax (805) 388-3433; www.laube.com.

EPCOS (formerly Siemens) Offers New Publication: Aluminum Electrolytic Capacitors Short Form Catalog 2000

EPCOS, Inc. has published *Aluminum Electrolytic Capacitors Short Form Catalog 2000* .

Aluminum electrolytic capacitors by EPCOS are used in a wide range of applications, including converter/traction, flashlights, lamp ballasts, SMPS (switched-mode power supply) and automotive. This short form catalog provides information at a glance about rated capacitance, quality grades, special features, temperature range and

applications. Products featured include screw terminal, soldering pin and 4-pin snap-in, flashlight capacitors, soldering star and axial, 3-pin terminal, snap-in terminal and new single-ended aluminum electrolytics.

This short form catalog may be viewed on the World Wide Web at www.epcos.com.

For further information or a copy of *Aluminum Electrolytic Capacitors Short Form Catalog 2000* , contact EPCOS, Inc. at 1-800-888-7729.

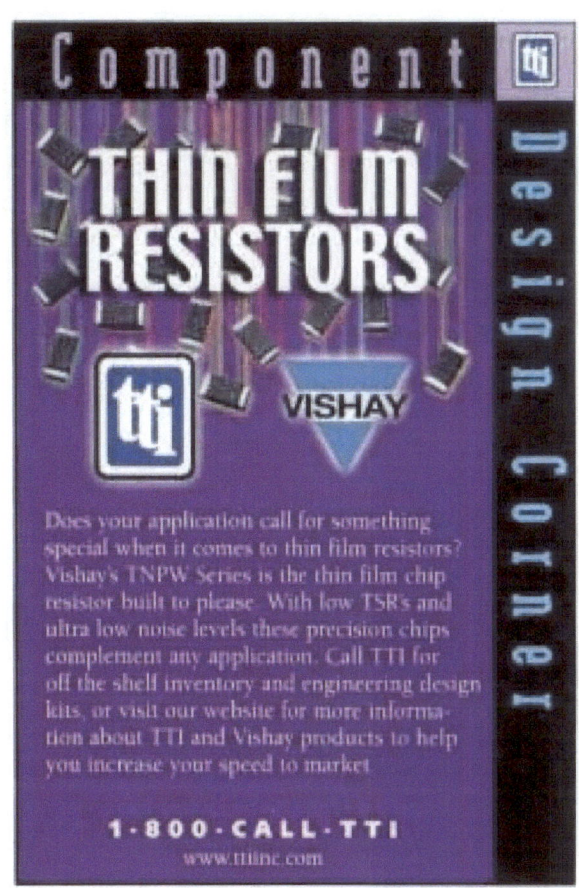

Powercache Expands Product Line with PC10 Ultracapacitor

PowerCache, a Maxwell Technologies company, announced the availability of the PC10, a miniature 10-farad ultracapacitor. Priced aggressively in volume purchases, the PC10 can increase battery operating life while reducing battery size in wireless consumer electronics and medical devices, automatic meter readers, scanners, power tools, or any other product requiring pulsed power capabilities. The PC10 can also be used in automotive subsystem applications for power locks, inside lighting and power windows.

Measuring 24 mm x 31 mm x 4.5 mm, the PC10 caches 31 joules of energy at a nominal 2.5 volts for high-powered discharges ranging from fractions of a second to one minute. According to Robert Tressler, Vice President of Sales and Marketing for PowerCache, its

small size and pulsed power capacity can decrease the weight and cost of batteries in devices requiring bursts of power.

"With roughly 10 times the power density of ordinary batteries, the PC10 can provide power during an application's peak periods, such as when sending messages or linking with satellite communications," said Tressler. "By doing so, the PC10 relieves batteries of peak power functions, so a product's life can be drastically extended while the overall system cost is simultaneously reduced."

In addition, the PC10 can provide extended backup power availability, allowing critical information and functions to remain available during dips, sags, and outages in the main power supply or battery charge. It can also be incorporated into automobile design. Once embedded into doors, the PC10 is trickle-charged through point-to-point contact and can provide any required power for locking, inside lighting and power window functions.

Available immediately, the PC10 is packaged in a durable, lightweight, hermetically-sealed stainless steel prismatic can. It features easily accessible, solderable terminals and an electrostatic storage capability that can cycle hundreds of thousands of charges and discharges without performance degradation. Like all PowerCache products, the PC10 is capable of accepting charges at the identical rate of discharges for systems that can benefit from regenerative energy.

PowerCache designs, develops and manufactures a family of large- and small-cell ultracapacitors. Its products are used in the power quality assurance, wireless communications, automotive, industrial automation, medical, personal digital assistant, actuator, automated meter reader, scanner and consumer electronic markets. The company's ultracapacitors extend the life of products by providing peak power requirements for overall energy management.

PowerCache is a part of Maxwell Technologies' Electronic Components Group, a family of subsidiaries that designs and manufactures electronic components for a variety of markets. Working with StratoTech Corporation, a world leader in designing and implementing high efficiency manufacturing methods, these subsidiaries operate demand-based manufacturing facilities.

The Electronic Components Group consists of PowerCache, Space Electronics, Inc., and Sierra-KD Components. PowerCache provides double-layer carbon ultracapacitors for applications requiring high pulses of power; Space Electronics provides radiation-hardened electronic components and board-level products for space and satellite applications; and Sierra-KD provides multilayer ceramic capacitors and EMI filters for aerospace, medical and high-grade industrial applications.

Maxwell Technologies applies industry-leading capabilities in power and computing to develop and market products and services for customers in multiple industries, including telecommunications, consumer electronics, satellite, energy, transportation, medical products, and water purification.

Sales contact: Bobby Maher, PowerCache Ultracapacitors, 4949 Greencraig Lane, San Diego, CA, 92123; + 1 (858) 576-7733; fax: + 1 (858) 503-5221; email: bmaher@maxwell.com. Visit our Web site at www.powercache.com.

Chip Scale Package Reduces Board Space Required for ESD Protection

California Micro Devices Corporation (CAMD) announced ESD protection products in Chip Scale Packaging (CSP). Two new devices, the PAC™DN1408C and

Continued on page 38

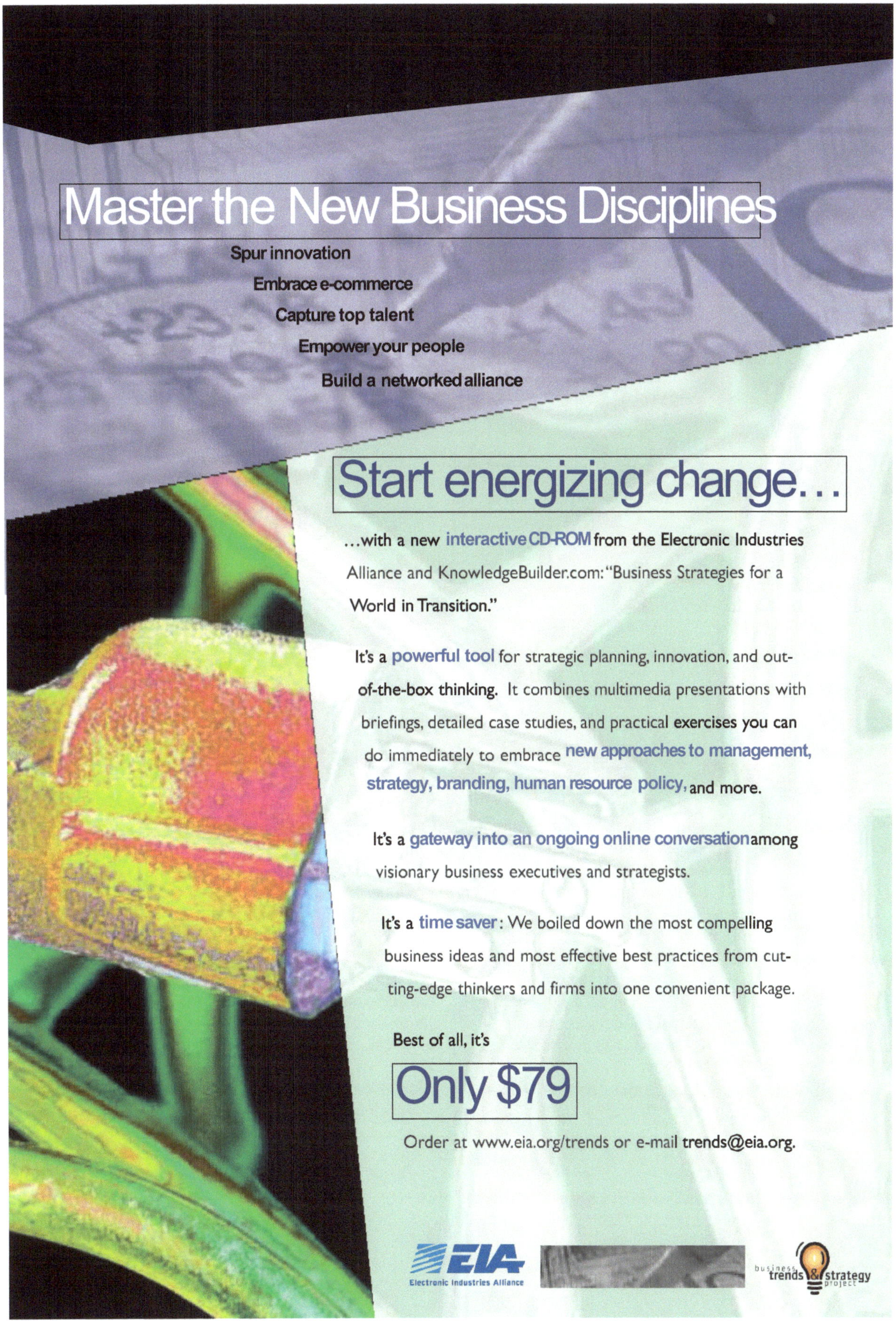

Master the New Business Disciplines

Spur innovation

Embrace e-commerce

Capture top talent

Empower your people

Build a networked alliance

Start energizing change…

…with a new interactive CD-ROM from the Electronic Industries Alliance and KnowledgeBuilder.com: "Business Strategies for a World in Transition."

It's a powerful tool for strategic planning, innovation, and out-of-the-box thinking. It combines multimedia presentations with briefings, detailed case studies, and practical exercises you can do immediately to embrace new approaches to management, strategy, branding, human resource policy, and more.

It's a gateway into an ongoing online conversation among visionary business executives and strategists.

It's a time saver: We boiled down the most compelling business ideas and most effective best practices from cutting-edge thinkers and firms into one convenient package.

Best of all, it's

Only $79

Order at www.eia.org/trends or e-mail trends@eia.org.

Continued from page 36

PAC™DN2408C, are CAMD's first ESD protection products available in the CSP format. The solder bumps allow attachment to laminate boards (such as industry standard FR-4 material) without the use of underfill. They are intended for use in applications where an extremely small foot print is required, especially cellular phones and PDAs, and for Internet appliances and PC ports where space savings might be desired. The PACDN2408C is configured with back-to-back zener diode connection s for protection of AC signal lines, such as analog line-level audio and video signals foun d in set-top boxes, DVD players and VCRs.

The pioneering use of chip scale packaging, which replaces the traditional "die on leadframe in molded plastic encapsulation," is a new standard product area for CAMD. Other ESD products from CAMD have been produced with surface mount format s for many years, but the new CSP format allows the assembly of these devices directly to industry standard FR-4 fiberglass laminate boards using existing surface mount technol- ogy production equipment. The use of large diameter solder balls com pensates for differential thermal ex- pansion, and underfill is not required. The CSP devices occu py less than twenty percent of the area of the con- ventional SOT and MSOP packages that they replace.

CAMD will add new configuration s to the ESD product family following this initial offering of eight-channel devices. CAMD continues to offer conventional pack- ages (such as SOT and MSOP) as part of the ESD prod- uct family.

All electronic products are susceptible to ESD damage from user or assembly generated discharges, or from "hot-plugging" of electronic systems. Advanced technology semicon ductors operating at low voltages are particularly sensitive. CAMD's ESD protection devices use proven technology to protect com ponents by absorbing or diverting threatening energy from ESD strikes. These two new devices are easy to use and are placed on board assemblies the same as other surface mount devices.

Pricing for both the PACDN 1408C and PACDN2408C in 10 K quantities is $0.35. Samples are available now. Production time is eight weeks ARO.

Full Line of SMT Carrier Products Offered by Cornett Taping Products

Cornett T aping Products, a leading manufacturer of embossed carrier tape, is now offering a broad line of cos t-effective SMT packaging products and services for OEMs.

Cornett Taping Products' SMT carrier tape is manufactured from a conductive polystyrene material and is available in widths ranging from 8 mm to 152 mm with thickness up to 24 mil. This material offers excellent static resistance and deep pocket capability up to 16 mm. Its surface provides superior adhesion and peel-back proper- ties for cover tapes.

The resultant product is a tough, protective, transport media designed for delicate components and precision devices. Nonconductive materials are available with anti-reflective and pedestal base designs. Pocket Ao, Bo, Ko, K1 dimensions are certified to ±.005 mm; each order has precision dimension certifications included.

Working from customers' specifications or product samples, Cornett Taping Products' application engineers can provide for your custom tooling needs. They have the capability to design to exacting tolerances for stan- dard or nonstandard devices, including custom pocket tooling and pedestal bottom pockets. Whether standard or custom designed, Cornett Taping Products' meet or exceed ANSI and EIA standards.

Besides carrier tape, Cornett Taping Products also supplies cover tape, reels, ESD supplies and bags. Quick response time and excellent customer service is guaranteed on all RFQs and requests for product information.

For a dditional information on Cornett Taping Products, visit their Web site at www.cornetttapingprod - ucts.com or contact: Ms. Colleen Nilsen, Vice President, Cornett Taping Products, 4221 Brickell Street, Ontario, CA, 91761. Toll free: (877) 650-TAPE; tel: (909) 937-9045; fax: (909) 937-9046.

Continued on page 40

Never Underestimate the Importance of Good Information

The Paumanok Group
Services for the Passive Component Industry

- Mergers & Acquistions
- OEM & CEM Pricing Strategies
- Strategies for Growth & Success
- Competitive Analysis

- Single-Client Market Research
- Multiclient Studies
- Conferences & Seminars
- On-Site Presentations

The Paumanok Group • 109 Kilm ayne Drive, Suite A • Cary, NC 27511 • USA
(919) 468-0384
(919) 468-0386 Fax
www.paumanokgroup.com
info@paumanokgroup.com

Newsmakers

Continued from page 38

RCD Offers Worlds Smallest Thin-Film Surface Mount Resistor

RCD Components Inc., a leading manufacturer of resistors, coils, and delay lines, has announced the release of their 0201 size thin-film surface mount resistor. RCD's new chip measures only 0.02 long x 0.01 wide, designated BLU0201 (Ultra-Precision Thin-Film) and is the smallest known thin-film in the industry. "RCD un- derstands the need for smaller ultra-precision surface mounts; until now the industry was forced to use much larger sizes," said Al Arcidy, VP Marketing. "We're now able to offer ultra-precision in the smallest resistor sizes; formerly this size was only available in a 'less pre- cise' thick-film variety," Arcidy added.

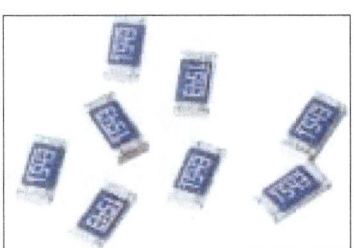
(Enlarged from actual size)

RCD's type BLU0201 offers a performance level unattainable in thick-film chips, designed specifically for demanding precision surface mount applications. Availability in a wide resistance range from 33 Ω to 22 Ω, and tolerances from 0.1% to 5% make this part suitable for almost all applications. Combining RCD's expertise in the field of ultra-precision resistors with the latest in automated c hip resistor production enables their pricing to be com parable with conventional preci- sion leaded resistors. Typical pricing is less than $.25 each in production volumes.

For samples or detailed product information, contact RCD Components Inc., at 520 East Industrial Park Drive, Manchester, NH, USA, 03109. Phone: (603) 669-0054; toll free order hot line: (877) RCD-COMP; fax: (603) 669-5455. The specifications for this product may also be accessed at www.red-comp.com or by dialing RCD's "Fax on Demand" service at (603) 669-0054, ext. 602 (Document #119).

NML Series: Multilayer Chip Inductors

NML series of multilayer chip inductors feature high Q (>50 at 1 GHz; <27 nH) and SRF (>2 GHz; <18 nH) for high frequency, wireless communication applications. Small 0402 and 0603 size  packages are ideal for handheld and ultra-dense circuit designs. Inductance values from 1.0 nH to 56 nH (0402) and 1.5 nH to 220 nH (0603) in ±0.5 nH, ±5% and ±10% tolerance; operating temperature range of -40° C to +85° C; supplied on paper tape and com patible with auto pick-and-placement and reflow soldering. Pricing ranges from $0.06 to $0.08 each in production volumes. Current lead time is six to 10 weeks.

Contact: NIC Components Corp., 70 Maxess Road, Melville, NY, 11747. Phone: (631) 396-7500; fax: (631) 396-7575; www.niccomp.com.

Jaro Announces New Lead-Free Chip Resistors That Meet Future 2005 Requirements

ISO-certified Jaro Components, Inc., a worldwide manufacturer of passive com ponents and thermal systems, has just announced a new lead-free, chip resistor —ju st one month after the introduction of a new lead-free, high-frequency inductor.

All of Jaro's lead-free products are designed to meet the upcoming 2005 lead-free requirement, as well as mandatory environmental protection standards that will soon require environmental impact substances such as lead to be eliminated. For com panies doing busi- ness in Europe, where the sale of lead-bearing electron- ic products will be restricted after January 1, 2004, this is a big plus.

Accor ding to Jaro spokesperson Dennis Eisen, "In ad- dition to improved environmental effect s, the elimination of lead from our resistors improves the heat resistance property for reflow (paste) soldering. This allows higher melting points for the solder to significantly shrink the process window."

These resistors are compatible with all soldering processes. Their resistance ranges from 1 ohm to 10 Mohm and is available in the following sizes: 0402, 0603, 0805, 1206, 1210, 2010 and 2512. Tolerances are avail- able at 1% or 5% with the following wattage ratings: 1/16 W, 1/10 W, 1/8 W, 1/4 W, 1/3 W, 3/4 W and 1 watt. The RC series comes in EIA Standard E96 and E24 resis- tance values.

These com ponents are ideal for hand-held devices and all other resistor applications where size is impor- tant. Pricing for the new lead-free chip resistors are $.007 each at 50,000 pieces. Delivery runs eight to 10 weeks ARO, and the product is packaged in 10 K, 25 K, and 50 K piece reels.

Specifications can be downloaded at www.jaro1.com/rc_specs.pdf . For more information about Jaro Components, Inc . and its growing list of cut- ting-edge products, contact Dennis Eisen at 1-561-241-6700 (x307), at dennis@jaro1.com or via the Web site at www.jaro1.com. ❑

ELIMINATE PALLADIUM

Base metal electrode solutions with X7R and Y5V BME dielectric compositions, copper end terminations, nickel inner electrodes and clean burn binder systems.

Ultra low-fire solutions with state-of-the-art COG and X7R dielectrics, matched internal electrodes, low-stress platable terminations and two-stage tape casting binder.

REDUCE PALLADIUM

ENHANCE MLV'S

Customized, multi-layered varistor solutions featuring platinum or palladium / silver electrodes, platable or solderable terminations and glass or metal oxide additives.

Multi-layered inductor solutions featuring non-magnetic core compositions, low DC resistance silver electrodes, platable silver terminations and wet stack / tape cast binder.

HIGH Q MLI'S

Even passive components require active technologies.

Technologies just like those found in Ferro materials. Technologies that improve performance. Technologies that increase production. Technologies that enhance your bottom line. We've got the goods you need to succeed. To learn more, just pick up your phone and give us a call. We share your passion for passive components.

Please reference AdSource #078PC when you inquire.

FERRO ELECTRONIC MATERIALS

Telephone: 760/305-1000
E-mail: adsource@ferro.com
Internet: www.ferro.com

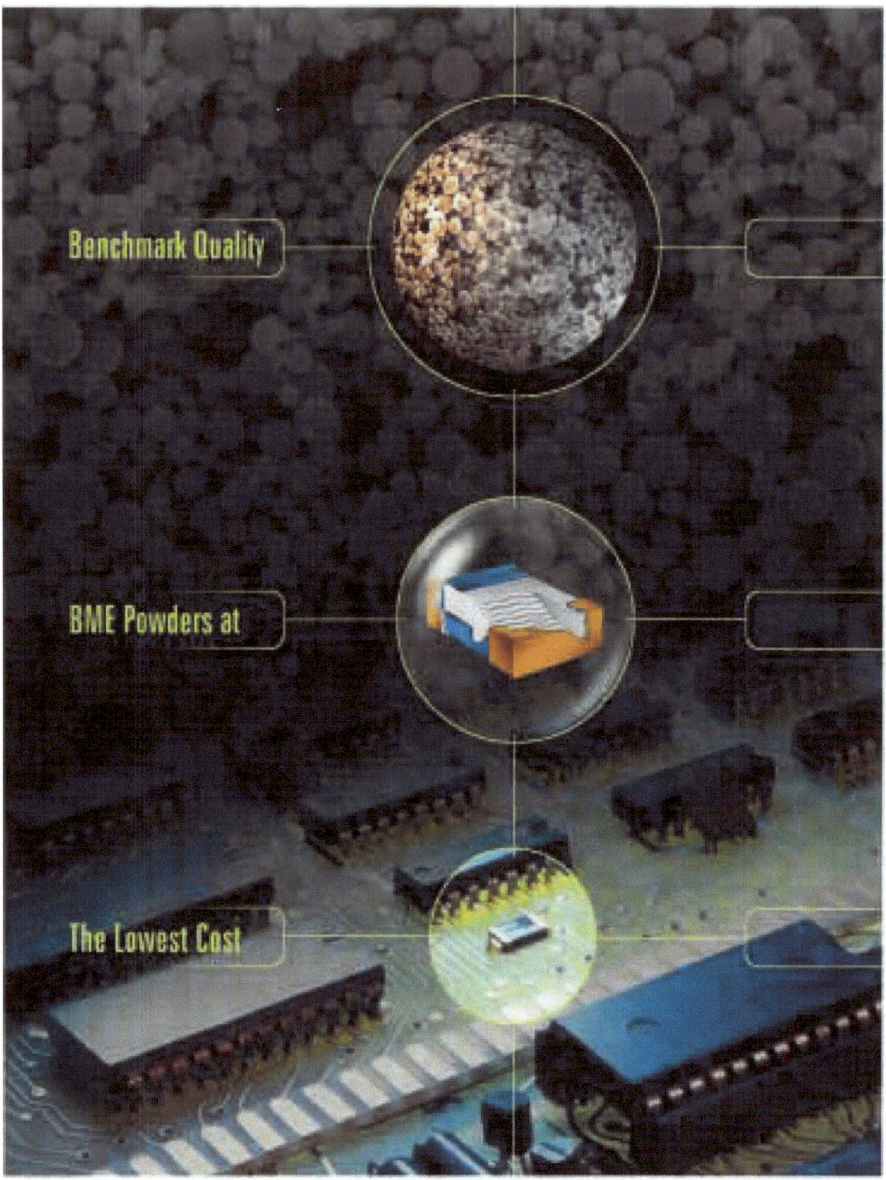

MAKING YOUR TRANSFORMATION TO BASE METAL SMOOTHER

CEPC is in the business of producing fine nickel & copper powder for the manufacture of base metal multi layered chip capacitors (BME-MLCC). With its high performance powders and technical expertise, CEPC is ready to team-up with your professionals to make your transformation from precious metals to base metals smoother. Thinner electrodes, more layers, higher capacitance are now within reach with our <0.3 μm powders. In association with our partners H.C. Stark, we have established World class teams to serve the North American, Asian and European MLCC industry through our joint distribution and customer assistance centers located in the U.S.A., Japan & Germany. Using high purity feed materials and a proprietary vapor phase technology that allows the controlled nucleation and growth of particles, CEPC's production process offers the unique ability to tailor the powder properties to meet even the most stringent demands of BME-MLCC manufacturers. These properties include degree of crystallinity, spherical particle shape and a mean diameter ranging from 0.1 - 1.5 μm. The spherical and highly crystalline nature of CEPC's powders make them highly resistant to oxidation and shrinkage. For more information, please contact us.

www.ingramcontent.com/pod-product-compliance
Lightning Source LLC
Chambersburg PA
CBHW051103180526
45172CB00002B/762